Cambridge Elements ≡

Elements in the Philosophy of Biology
edited by
Grant Ramsey
KU Leuven
Michael Ruse
Florida State University

MECHANISMS IN MOLECULAR BIOLOGY

Tudor M. Baetu
Université du Québec à Trois-Rivières

CAMBRIDGE
UNIVERSITY PRESS

CAMBRIDGE
UNIVERSITY PRESS

University Printing House, Cambridge CB2 8BS, United Kingdom

One Liberty Plaza, 20th Floor, New York, NY 10006, USA

477 Williamstown Road, Port Melbourne, VIC 3207, Australia

314–321, 3rd Floor, Plot 3, Splendor Forum, Jasola District Centre, New Delhi – 110025, India

79 Anson Road, #06–04/06, Singapore 079906

Cambridge University Press is part of the University of Cambridge.

It furthers the University's mission by disseminating knowledge in the pursuit of education, learning, and research at the highest international levels of excellence.

www.cambridge.org
Information on this title: www.cambridge.org/9781108742306
DOI: 10.1017/9781108592925

First published 2019

A catalogue record for this publication is available from the British Library.

ISBN 978-1-108-74230-6 Paperback
2515-1126 (online)
2515-1118 (print)

Mechanisms in Molecular Biology

Elements in the Philosophy of Biology

DOI: 10.1017/9781108592925
First published online: September 2019

Tudor M. Baetu
Université du Québec à Trois-Rivières
Author for correspondence: Tudor M. Baetu, tudor-mihai.baetu@uqtr.ca

Abstract: The new mechanistic philosophy is divided into two largely disconnected projects. One deals with a metaphysical inquiry into how mechanisms relate to issues such as causation, capacities, and levels of organization, while the other deals with epistemic issues related to the discovery of mechanisms and the intelligibility of mechanistic representations. Tudor M. Baetu explores and explains these projects, and shows how the gap between them can be bridged. His proposed account is compatible both with the assumptions and practices of experimental design in biological research, and with scientifically accepted interpretations of experimental results.

Keywords: mechanism, new mechanistic philosophy, biology, philosophy of science, philosophy of biology

ISBNs: 9781108742306 (PB), 9781108592925 (OC)
ISSNs: 2515-1126 (online), 2515-1118 (print)

Contents

1 Mechanisms and Their Discovery

1.1 Mechanisms and Mechanistic Explanations

Many explanations in biology tell us how phenomena are produced as a result of changes and interactions among the various parts of a physical system. Such systems of parts changing over time are commonly known as 'mechanisms', while the more-or-less detailed descriptions of these systems and the series of changes they undergo are referred to as 'mechanistic explanations'.

If the reader is left wondering what a phenomenon is and how a mechanism explains it, a concrete example may help. Biologists take reproducible sequences of events, whereby exposure of a biological system to a stimulus is consistently followed by a similar kind of response, to be phenomena in need of an explanation (Figure 1). Take sunburns, for example. Prolonged exposure to ultraviolet radiation consistently results in a pronounced redness and swelling of the skin. Since this sequence of events can be consistently reproduced in humans and many other animals, it is unlikely to be just a coincidence. It is something that demands an explanation.

Sunburns belong to a family of related phenomena of great clinical importance known as inflammatory responses. These responses are triggered by a wide variety of harmful stimuli, including radiation, tissue damage, pathogens, and antigens, and typically consist of five directly observable symptoms: erythema (redness), edema (swelling), pain, heat, and loss of function (e.g., immobility). In most cases, inflammatory responses shut down after the threat has been eliminated or the harmful stimulation has subsided, which is highly desirable since a prolonged or chronic inflammatory response has detrimental consequences for the organism.

Much research in immunology and medicine has been devoted to finding out how – that is, by means of which mechanisms – inflammatory responses are produced. Since my goal is simply to give the reader an idea of what a mechanistic explanation may look like, I will focus on one of the better understood mechanisms involving a type of white blood cells known as T-cells. By the end of the 1980s, it had been established that, when an organism is exposed to harmful stimuli, T-cells begin secreting a variety of molecules necessary for mounting inflammatory responses. Thus, one key research question was to find out how T-cells are activated. The molecular underpinnings of T-cell activation turned out to revolve around a couple of regulatory DNA-binding proteins, one of which is the transcriptional factor nuclear factor κB, or NF-κB in short (Sun et al. 1993). In resting T-cells, NF-κB is trapped in the cytoplasm by a protein known as inhibitor of NF-κB, or IκB. When cells are exposed to harmful stimuli (Figure 2A), however, a chain of protein–protein interactions

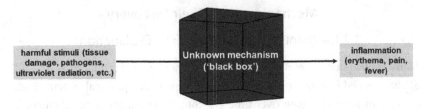

Figure 1 Inflammatory responses as phenomena to be explained

leads to the dissociation of inactive IκB/NF-κB complexes (B). NF-κB is freed (C) and can now move to the nucleus, where it binds specific DNA sequences and enhances the expression of a number of genes, many involved in immune responses (D). This results in a variety of new proteins being manufactured in the cell, such as the cyclooxygenase-2 (COX-2) enzyme, which catalyzes the synthesis of prostaglandins. Once secreted in the bloodstream, prostaglandins promote vasodilation, which is responsible for erythema and edema (redness and swelling of the skin); sensitize spinal neurons to pain; and act on the thermoregulatory center of the hypothalamus to produce fever (E).

We can now form a general idea about how inflammatory responses are initiated when organisms are exposed to harmful stimuli. What is missing is an explanation of how the responses shut down. It was eventually discovered that NF-κB also enhances the production of its own inhibitor, IκB. The newly synthesized IκB binds NF-κB, trapping it back in the cytoplasm (F). Prostaglandins too curb NF-κB activity by interfering with the signaling pathway leading to the degradation of the inhibitory protein IκB (G). Thus, following stimulation, T-cells are initially activated, resulting in an inflammatory response, then are automatically turned off by means of a negative feedback loop molecular mechanism, which performs a function analogous to that of a common thermostat.

The above explanation is in many respects incomplete. For instance, it is not specified how prostaglandins cause the symptoms of inflammation. Quantitative-dynamic details are also missing. The explanation doesn't say anything about the duration of the inflammatory response or whether cyclical inflammatory responses are generated in response to persistent stimulation. Finally, some familiarity with chemistry is tacitly assumed, such as the notion that protein–protein and protein–DNA interactions are weaker forms of binding in comparison to the covalent bonds holding molecules together. Nevertheless, despite assumed, abstracted, and unknown details, the above narrative and the accompanying diagram should help the reader imagine how inflammatory responses are produced as a result of a sequence of changes and interactions involving various cellular and molecular components of an organism.

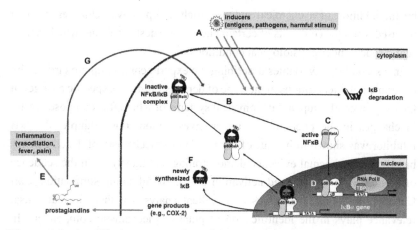

Figure 2 A simplified representation of the NF-κB mechanism and its role in regulating inflammatory responses

1.2 The Discovery of Biological Mechanisms

The elucidation of biological mechanisms often begins with the formulation of hypotheses sketching out possible mechanisms.[1] For instance, experimental results indicating that protein synthesis inhibitors block T-cell activation led some researchers to hypothesize that inflammatory responses rely on a mechanism of genome expression regulation. Initial speculative work is followed by what Lindley Darden (2006b, Ch. 4) describes as a gradual filling in of missing mechanistic details. This usually requires a significant amount of experimental research.

In a first step, a system responsible for producing the phenomenon of interest, or what William Bechtel and Robert Richardson (2010, Ch. 3) call the 'locus of control', is identified. In practice, this amounts to the characterization of an experimental setup in the context of which a phenomenon can be consistently reproduced. For instance, the experimental setup used to elucidate the NF-κB regulatory mechanism consisted mainly of a cell model of T-cell activation. Normal T-cells extracted from the blood of a healthy donor or precancerous ('immortalized') T-cell lines were grown in an artificial medium and stimulated

---

[1] Peter Machamer, Lindley Darden, and Carl Craver (2000, 18) define a mechanism sketch as "an abstraction for which bottom out entities and activities cannot (yet) be supplied or which contains gaps in its stages. The productive continuity from one stage to the next has missing pieces, black boxes, which we do not yet know how to fill in. A sketch thus serves to indicate what further work needs to be done in order to have a mechanism schema." In contrast, a more detailed mechanism schema, such as the one depicted in Figure 2, is a "truncated abstract description of a mechanism that can be filled with descriptions of known component parts and activities ... When instantiated, mechanism schemata yield mechanistic explanations of the phenomenon that the mechanism produces" (2000, 15, 17).

by the addition of a chemical inducer, such as lipopolysaccharides, a toxin released when the cell walls of certain bacteria are destroyed, and which causes septic shock under physiological conditions.

In a second step, variables describing the experimental setup are targeted by experimental interventions in the hope of demonstrating that specific changes in the experimental setup and the physical systems of which it is composed result in changes in the phenomenon under investigation. For example, the IκB inhibitor was shown to be part of the regulatory mechanism of T-cell activity based on experimental evidence demonstrating that mutations in the sequence of IκB result in a prolonged activation of T-cells following stimulation. Carl Craver (2007, Chs. 2–3) gives an excellent analysis of the role that causal relevance plays in the identification of putative mechanistic components. In particular, he points out that causal relevance is demonstrated by means of controlled experiments, thus establishing a connection between the scientific practices involved in the elucidation of mechanisms and James Woodward's (2003) interventionist account of causation.

Finally, a mechanism must eventually be 'recomposed' in order to show how it generates the phenomenon (Bechtel 2011). A mechanism's recomposition can be physical, for instance, an in vitro reconstitution experiment, or conceptual, such as a computer simulation or simply a narrative or diagram such as those illustrated earlier. One way or another, the goal is to demonstrate that components organized, acting, and having the properties described in the mechanistic explanation can produce, and ideally are sufficient to produce the phenomenon under investigation.

1.3 Experimental Methodology First

In scientific practice, hypothesis and experimentation proceed in tandem. The initial characterization of an experimental setup and the choice of variables targeted for intervention are motivated by existing hypotheses about the mechanism responsible for a phenomenon. Conversely, experimental results, such as the ability or failure to consistently reproduce the phenomenon, as well as the fortuitous discovery of causally relevant factors, play a key role in the formulation of new mechanistic hypotheses.

This Element, however, focuses almost entirely on experimental inquiry.[2] There are two reasons for this unusual choice. The first has to do with the fact that most philosophers approach experimental practice from the perspective of

---

[2] Perhaps a clarification is needed here. What I mean by 'experimental' or 'empirical inquiry' is experimental research in science. Experimentalists approach hypothetical explanations with an attitude of 'suspended belief' until adequate supporting evidence is produced. Being cautious, however, is not the same as being skeptical. Experimental research is a sustained effort to produce

famous discovery episodes that have left a profound mark on the development of science. Major discoveries are brought under philosophical scrutiny in an attempt to reconstruct the reasoning involved in the formulation of particularly fruitful hypotheses and the design and execution of famous experiments testing such hypotheses. Lindley Darden (1991; 2006b), William Bechtel (2006; 2008), Carl Craver (2007), and Marcel Weber (2005) analyze specific cases from biochemistry, genetics, cell biology, molecular biology, electrophysiology, and neuroscience, examining in detail some of the most famous discoveries of mechanisms in biology. Nevertheless, science can also be approached from the less glamorous, but equally relevant perspective of the basic methodological principles governing everyday 'normal science' research, as Kuhn would put it. The latter crystallized over the past four centuries into an autonomous set of practices governing experimental design independently of any specific explanatory hypotheses. Thus, one rationale for approaching mechanistic discovery from the standpoint of experimental methodology is that it has the potential of providing a novel and, I think, more generally applicable perspective on mechanisms and their discovery.

The second rationale is intimately linked to the immediate goal of this Element, which is to elaborate a metaphysical account of mechanisms. A number of important publications have recently brought mechanistic metaphysics under systematic scrutiny, most notably Stuart Glennan's (2017) *The New Mechanical Philosophy*. Nevertheless, these works also reveal a disconnect between two lines of philosophical inquiry. One deals with epistemic issues related to the discovery of biological mechanisms and the intelligibility of mechanistic representations. The other relates to a metaphysical inquiry into how mechanisms relate to issues such as ontology, causation, laws, and levels of organization. The main object of interest here is no longer discovery, but rather the final product of scientific research, namely the narratives and diagrammatic representations typically found in biology textbooks, which are analyzed in order to gain insights into what these epistemic products presuppose and entail from a metaphysical point of view. Despite the individual contributions of each project, the two remain disconnected inasmuch as it is not clear how the relatively modest and down-to-earth discovery strategies identified by the epistemic project can justify the more audacious claims associated with the metaphysical project.

One way to bridge the gap between the two projects is to stipulate that a metaphysical account of biological mechanisms should remain compatible

---

evidence for or against explanatory accounts, which is a clear indication that empirical inquiry in science doesn't condone the defeatist attitude typically associated with philosophical empiricism.

with the ontological assumptions of the experimental methodology employed in the elucidation of mechanisms. I think that the immense success of experimental research in biology – and it must be emphasized that biology is an experimental discipline to a much larger extent than physics and chemistry – justifies this demand. I argue therefore that two kinds of considerations should constrain mechanistic metaphysics. The more fundamental ones, amounting to what I call the 'minimal experimental interpretation', are methodological in nature. Two-thirds of this Element is devoted to these considerations. I begin my inquiry with an attempt to define the notion of 'phenomenon', which, surprisingly, received little attention in the mechanistic literature. In Section 2, I defend the view that standard experimental methodology assumes a causal interpretation of measurements in virtue of which causes responsible for differences in measured values of variables can be localized within the spatiotemporal boundaries of physical systems satisfying a given experimental description. This interpretation allows for a definition of phenomena in strictly methodological terms, as data reproduced when experiments are replicated.

An account of phenomena further determines how one construes the relationship between mechanisms and phenomena. In Section 3, I argue that neither of the two main philosophical accounts, the etiological and the part–whole constitutive accounts, is compatible with the demands of experimental research. I reject both accounts in favor of an alternative one according to which the mechanism responsible for a phenomenon is a causal structure that does not allow the variables probed by the measurements involved in the description of the phenomenon to vary independently of one another.

The foundational work conducted in Sections 2 and 3 is meant to provide a minimal framework onto which richer mechanistic ontologies, of the sort typically endorsed by scientists and philosophers, may be grafted. This brings us to the second kind of considerations constraining mechanistic metaphysics, namely those linked to a richer and more diverse set of physical interpretations relying on background knowledge about the physics and chemistry of biological systems. A considerable body of philosophical work on mechanisms targets, directly or indirectly, this kind of considerations. In the philosophical literature, mechanisms are systematically characterized as physical systems composed of spatiotemporally organized parts acting, interacting, or functioning in such a way as to produce, maintain, underlie, or constitute phenomena (Bechtel and Abrahamsen 2005; Glennan 1996; 2002; 2017, Ch. 2; Illari and Williamson 2012; Machamer et al. 2000). These characterizations depict mechanisms as consisting of physical entities acting or playing certain functional roles, or again as parts interacting in accordance to the laws of physics. In Section 4, I review the main interpretations circulating in the philosophical literature, along with their theoretical and experimental justifications. An exhaustive treatment of these interpretations is beyond the scope of this short Element. The main goals here are to make explicit the ontological commitments associated with each interpretation and to assess their compatibility with the minimal experimental interpretation developed in previous sections. My hope is that the analysis conducted in this section will enable the reader to gain a clearer understanding of how tacit and explicit metaphysical assumptions underpinning philosophical accounts of biological mechanisms relate to scientific theories and experimental practice. Finally, a recapitulation of the main theses defended, including some last-minute clarifications, is provided in the concluding section.

2 What Is a Phenomenon?

A mechanism is invariably characterized as a 'mechanism for a phenomenon'. But what is a phenomenon? Scientists describe phenomena as being both constitutive of empirical reality and objects of explanation, hinting to an overlap between these notions. Logical empiricists took this description at face value, equating phenomena, *explananda*, and observations. Some observations are unaided, and some are mediated by instruments of measurement, but they ultimately all amount to subjective perceptual experiences reported in the medium of language. My task in this section is to introduce an alternative view according to which empirical research in contemporary biology is primarily a matter of conducting controlled experiments in order to generate data structured in such a way as to make possible inferences about the causal structure of the world. Then I argue that, thus understood, data are phenomena to be explained.

2.1 Data

In order to understand what data are and how they relate to phenomena, it is useful to consider once again the example of inflammatory responses. As stated in the previous section, these responses are biological outcomes characterized by the occurrence of five symptoms – erythema, edema, pain, heat, and loss of function – following harmful stimulation, such as tissue damage or exposure to ultraviolet radiation. Based on this example, a phenomenon may be viewed as a 'black box' causal process initiated by a set of stimuli and terminating in a set of symptoms (Figure 1).

Each of the inputs and outputs of the phenomenon is treated as a variable whose values are given by measurements involving a specified technique. Some variables have a clear physical interpretation, which is to say that we know

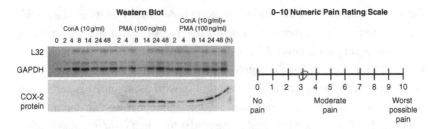

Figure 3 Data

something about the physical nature of that which is measured or, if we prefer, the factor to which the variable refers. Other variables don't have a physical interpretation. A Western blot detects proteins (Figure 3, left). In contrast, a pain assessment test measures pain, yet it is not clear what pain is beyond 'that which is reported on the occasion of a pain assessment test' (Figure 3, right). Erythema assessment stands somewhere in between. It measures several unspecified physiological changes, which include an increase of blood flow in the skin. As a rule, the more that is known about the causal structure of the world and the inner workings of the measurement technique, the clearer the physical interpretation. If little is known or if disputes arise about the correct interpretation, the only practicable option is operationalization. In this case, scientists distinguish and define different variables according to the techniques used to measure them, although it should be understood that there are no reasons to assume a one-to-one correspondence between operationalized variables and features of physical reality (Bridgman 1927, Ch. 1).

As a first approximation, data are measured values of variables. For instance, erythema is measured by visual assessment and assigned values ranging from 'no erythema' to 'violet erythema with edema'. In humans, pain intensity is measured by eliciting a physical response such as stating or pointing to a value on a numerical scale. In rats, pain is typically assessed by the Randall–Selitto test, which involves a measure of the threshold pressure at which the animal withdraws an inflamed paw. Body temperature values are given by thermometer readings. Loss of function assessment varies depending on the locus of inflammation. For instance, the severity of rheumatoid arthritis is measured by a mobility test, such as the fingertip-to-palm distance during maximal finger flexion.

While measurements do not preclude conventionalist elements, such as the choice of a particular scale, most experimental scientists assume that data consists of physical effects informative of the causal structure of the world (Trout 1998, 56–57). From a bare-bones methodological point of view, this simply means that differences in measured values reflect differences in the causal structure of the world. If this working hypothesis is correct, comparisons between measurements can be used to localize and identify causal difference-makers. The general strategy is to repeat measurements under circumstances that are kept constant in some respects and deliberately changed in other respects; the former are part of a 'control' condition, the latter of the 'test' condition.

For example, the attempts at standardization transparent in the description of the methods of assessment of erythema and pain intensity are an elementary form of experimental control. Measurement techniques are carefully replicated from one measurement to the next and designed in such a way as to generate perceptually unambiguous outputs meant to eliminate potential disagreements among observers. Standardization aims to ensure that any particular instantiation of the technique and any particular observer can be exchanged with any other without affecting the data. If such exchangeability can be assumed, it is safe to infer that differences in data outputs are due to reasons other than differences in the measurement system. Conversely, comparisons between data generated by different or unstandardized measurement techniques are inconclusive since it is not clear what is kept constant from one measurement to the next. Such comparisons may still be valid, but they require a theory specifying a relationship between variables, such as a common physical interpretation.

The pattern of reasoning illustrated above is an application of the venerable method of difference (Mill 1843, Chapter VIII, § 2), a type of contrastive inference whereby those aspects of a situation kept constant are ruled out as possible causes responsible for differences in effects. Ruling out certain factors is useful inasmuch as it enables researchers to make more-or-less precise claims about the physical localization of the causes responsible for differences in measurements (Bechtel and Richardson 2010). In the case of standardization, the universe is decomposed into two parts: the measurement system, which is assumed to consist of essentially identical copies, and the rest of the universe, a rather large chunk of physical reality containing a yet-to-be-circumscribed object of study. A contrastive inference allows researchers to conclude that, inasmuch as the measurement system is kept constant, the causes responsible for changes in the measured values of a variable lie outside the measurement system; in other words, that something 'out there' is indeed being measured.

2.2 Experimental Models of Phenomena

In more general terms, variations in data are attributed to variations of multiple causes at work in the particular circumstances of each measurement

(Fisher 1935).[3] Fixing these circumstances is expected to reduce variability. In most cases, this is indeed what happens. Data varies less as the spatio-temporal interval between measurements is reduced. A more interesting result, however, is that data obtained by replicating methods and experimental setups also tends to display less variation than data obtained when no such precautions are taken. In some cases, this is true even if measurements are repeated after a long period of time, at different locations, and by different researchers.

Any scientific paper presenting original experimental findings contains a materials and methods section in which techniques and experimental setups are carefully described with the explicit goal of increasing the prospect that subsequent experiments replicating the same methods and circumstances will reproduce similar data. When an experimental setup is described in enough relevant detail as to ensure that highly similar data are systematically obtained over a large number of replication experiments, it is often referred to as an *experimental model of a phenomenon*. For instance, an experimental model of inflammatory responses widely used in basic and clinical research is described as follows:

> UV radiation between 270 and 400 nm, peaking at 310 nm was delivered from 10 fluorescent UV-B lamps, Philips TL20 W/12 (Philips GmbH, Hamburg, Germany), housed in a UV 800 unit (Waldmann GmbH, VS-Schwenningen, Germany). UV-B irradiance (280–320 nm) at the surface of the test areas was measured with a calibrated radiometer equipped with a SCS 280 photodetector (International Light, Newburyport, Mass., USA), and was 2.4 mW/cm^2 at a tube to target a distance of 40 cm ... Erythema was determined by visual assessment 24 h after irradiation and was graded as follows: 0 = no erythema; 1 = barely perceptible erythema with sharp borders (MED); 2 = light red, marked erythema; 3 = dark red, marked erythema; 4 = violet erythema, edema. (Jocher et al. 2005)

The above procedure, known as the 'ultraviolet-induced erythema assay', consists in inducing artificial sunburns. The inducer of the inflammatory response, or harmful stimulus, is ultraviolet radiation, while a measure of the severity of erythema serves as a proxy for estimating the severity of the response. By specifying, measuring, and ultimately standardizing the inducer it eventually became possible to consistently reproduce inflammatory responses of a desired intensity.[4]

---

[3] For a historical overview of the development of statistics, see Hacking (1990) and Stigler (1986).

[4] For a history of the ultraviolet-induced erythema assay, see Brune and Hinz (2004). For a philosophical discussion of models, including experimental models, see Ankeny (2001), Ankeny and Leonelli (2011), and Baetu (2014).

The epistemic significance of the model is that it specifies a replicable and recognizable physical object of investigation, namely the experimental setup, likely to contain within its spatiotemporal boundaries all the causal ingredients required for producing an inflammatory response of the skin. The general pattern of inference at work here goes as follows: If the description of the experimental setup were such that it is satisfied both by physical structures that contain relevant causal factors and structures that don't, the data would have varied among the many measurement experiments. Since this is not the case, it may be concluded that a set of proximal causes sufficient for generating the same differences in measurements is stably associated with experimental setups satisfying this description, being present wherever and whenever the setup is relocated or reconstructed. A likely explanation of the stability of input–output associations is that the causes responsible for differences in data are intrinsic to the replicated experimental setup, possibly because they coincide with the referents of the labels used to characterize and replicate the setup, or because they are linked to these labels as cause and effect or as effects of a common causal background (e.g., a latent variable), or perhaps because they are inherent to the measurement and intervention techniques, in which case the data generated amount to experimental artefacts.

2.3 Reproducibility

In practice gains (or losses) in reproducibility function more as a heuristic guiding research than as a method for conclusively demonstrating localization. An inductive generalization whereby a finite number of successful reproductions are assumed to be representative of the infinitely many possible replication experiments is defeasible. Claims about reproducibility are revised as additional replication experiments are conducted and a larger body of data is produced. Data initially reproduced may turn out to be irreproducible as additional replication experiments are conducted, or vice versa. These simpler cases are amenable to statistical testing for false positives or negatives. In other, more challenging cases, data may turn out to display order in the form of stochastic biases over a large number of replications. Conversely, higher-resolution measurements conducted in the context of an experimental setup known to generate reproducible effects may reveal stochastic irregularities. The lytic–lysogenic cycle of the lambda bacteriophage illustrates the former (Ptashne 1998), while cellular and molecular noise are examples of the latter (Elowitz et al. 2002). Such cases are marked by a more drastic rethinking of what counts as the data set to be reproduced.

A more immediate worry is that data are never exactly reproduced in the first place.[5] James Bogen and James Woodward (1988) point out that data are inherently 'noisy', varying from one measurement to the next even in the trivial case when consecutive measurements are taken in the context of the same experimental setup. This presents a rather unappealing choice: either accept that data are never reproduced and their causes ultimately never localized and identified, as Bogen and Woodward themselves conclude, or, following a path suggested by James McAllister (1997), rescue reproducibility by drawing a distinction between significant and insignificant data variation. The latter solves the problem by introducing a form of conventionalism according to which phenomena are investigator-relative constructs. Ultimately, this entails that the boundaries of mechanisms too cannot be objectively fixed, but shift depending on what individual researchers or the scientific community consider to be worthwhile *explananda*.

There is, however, a third option that not only avoids these extremes, but also best describes experimental practice. From an experimental standpoint, reproducibility failures can be either

(a) variations due to insufficiently stringent descriptions capturing a wide variety of physical structures superficially similar at the level of the control labels used to describe and replicate experimental setups; or

(b) variations independent of experimental descriptions, such as the inherently stochastic nature of certain causal processes, causal heterogeneity inherent to natural populations of biological systems, and unavoidable coincidental interferences.

Type (a) failures are dependent on the choice of the experimental setup and therefore can be addressed by increasing experimental control. They are often tackled by trial-and-error attempts to develop new experimental setups or tweak existing ones. Gains in reproducibility can be attributed to the fact that more-stringent descriptions are no longer satisfied by causally irrelevant physical structures captured under previous descriptions, thus increasing the likelihood that the relevant causal factors are associated with the experimental setup on the occasion of a particular replication experiment. Type (b) failures, on the other

---

[5] This is in part because it is not always possible or desirable to exactly replicate experimental setups. For example, it is usually impossible to replicate an experiment using the same patients. Practicalities often dictate that descriptive labels and standardization criteria deemed likely to be important are replicated with as little variation in their values as possible, while more variation is allowed in other respects. Robustness to variation also secures some degree of generality. For example, there is little incentive in exactly replicating an experiment demonstrating the same outcomes for a specific patient; from a clinical perspective, it is much more relevant to reproduce more-or-less similar outcomes for a large number of patients.

hand, reflect a residual variation in data that persists despite all efforts to increase experimental control.

Common experimental strategies for increasing reproducibility include a more rigorous assessment of outcomes and the standardization of intervention techniques, increasing the number of control labels describing experimental setups, making a more judicious choice of labels based on theoretical expectations, and designing setups based on a partial or hypothetical understanding of the mechanisms underpinning the phenomenon of interest. Clinical research on inflammatory responses benefited from improved standardization. In the early decades of the twentieth century, researchers seeking to develop better treatments for rheumatoid arthritis, a medical condition characterized by an inflammation of the joints, faced a problem of poor reproducibility. The severity of rheumatoid arthritis varies greatly not only between patients, but also over time for the same patient. This background variability made it difficult to assess the causal efficacy of experimental and medical interventions since the effect of the intervention could not be disentangled from that of myriad other causal factors concomitantly impacting on the severity of the condition. Once the ultraviolet-induced erythema assay was developed, it became possible to reliably elicit local inflammatory responses of the desired severity. This experimental model provides a standardized control condition marking a constant 'base line' relative to which the outcomes of interventions can be interpreted in terms of causal relevance. Moreover, the widespread adoption of the same experimental model by different labs enabled researchers to compile findings from different sources into a coherent body of knowledge. For instance, this made possible comparisons between the relative magnitudes of the efficacy of various anti-inflammatory drugs. By contrast, if different labs had continued to rely on different experimental setups, such comparisons would have been meaningless.

The fact that reproducibility can be enhanced by modifying experimental setups demonstrates that data are not categorically 'clean' or 'noisy', reproducible or irreproducible, but rather come in degrees of noisiness and reproducibility, depending on the experimental setup used to generate them. This does not mean, however, that degrees of reproducibility are subjective constructs. Unlike McAllister's thresholds of statistical significance, which are imposed by convention, degrees of reproducibility are discovered a posteriori, by trial and error, and are objective, in the sense that they are determined by the physical structure of experimental setups.

It is also interesting to note that unless the extent of type (b) failures can be accurately predicted by an accepted theory, it cannot be decided a priori whether researchers should persist in trying to increase experimental control or accept data variability as unavoidable. Thus, there is always uncertainty about whether the maximum degree of experimental control and reproducibility has been attained, or whether researchers are held back by technological limitations, misleading background theories, or sheer bad luck. Historical case studies reveal that research is typically driven by a reciprocal feedback between attempts at explanation and experimental fiddling aimed at increasing reproducibility. Experimental tweaks are suggested by existing theories and beliefs. In turn, the survival of the latter depends to a large extent on whether they lead to gains in data reproducibility. Once some measurable degree of reproducibility is secured, the control labels on the basis of which experimental setups are replicated provide a preliminary list of causally relevant factors marking the first step in the elucidation of mechanisms. The eventual discovery of such factors leads to the formulation of new explanatory hypotheses, which in turn suggest further experimental refinements, thus perpetuating new cycles of explanation-experimentation. For example, the discovery of hemolysis and hemagglutination reactions when blood from different species is mixed together marked the first major breakthrough in increasing the safety of blood transfusions. In turn, the increase in the success rate of human-to-human over animal-to-human transfusions contributed to the dismissal of earlier theories drawing on a putative relationship between humors and personality, marking the beginning of a new explanatory paradigm that eventually led to the elucidation of the cellular and molecular mechanisms of humoral immunity (Schwarz and Dorner 2003).

2.4 Noise

Type (b) failures arise when attempts to increase experimental control are unsuccessful or undesirable. The most common example of unsuccessful attempts to increase experimental control are limitations in measurement precision. Clinical research, as well as research involving field studies, illustrates cases where an increase in experimental control is undesirable. For example, once it has been established that ibuprofen, a nonsteroidal antiinflammatory drug developed in the 1950s, consistently reduces ultraviolet-induced erythema in guinea pigs, the next step was to test the efficacy of the drug in a small-scale randomized controlled trial on patients with rheumatoid arthritis.[6] The first experimental setup is tightly controlled, with little variability in terms of the severity of inflammation and differences between test animals. The second experimental setup consists of samples drawn from a heterogeneous human population. Here, the antiinflammatory effect of the drug varies a great deal more, depending on which patients

---

[6] A historical account of the discovery of ibuprofen can be found in (Rainsford 2015).

happen to be enrolled in the trial and how severe their condition happens to be at the time of assessment. Yet, despite the marked loss in reproducibility, this second experiment is essential for demonstrating the efficacy of the drug in humans.

Variation that cannot or should not be reduced by increasing experimental control is commonly dealt with by statistical methods. It is common practice to take not just one measurement, but several measurements on each occasion and aggregate the resulting data as mean and standard uncertainty (Taylor 1997, Ch. 5). Likewise, contrastive inferences take the form of statistical tests whereby the variation of data in test and control conditions is used to estimate the probability that the two data sets originate from the same population of measurements (Fisher 1935; Hill 1955). From a physical point of view, standard statistical analysis treats type (b) variation as unwanted noise. The causal interpretation of data still applies, and noise is in principle explainable. It is just that this explanation is not particularly interesting. In the case of a clinical trial, the proximal cause of data variation is the sampling of patients and their random assignment to the test and control groups. In the case of measurement precision, variation is attributed to coincidental processes that are not of immediate interest to the researchers.

It may be tempting to conclude that the physical interpretation of type (b) variation is largely a matter of conjecture, which eventually solidified into convention as statistics gained acceptance among experimental researchers.[7] While this may be true in some contexts, this shouldn't overshadow the fact that a statistical model embodies assumptions about causal factors impacting on data, some of which can be independently tested. Moreover, biology offers examples of unavoidable variability other than measurement or sampling error. From a mechanistic perspective, two competing interpretations are available. One states that mechanisms fulfill their biologically significant activities, such as produce certain outcomes in response to stimuli, despite internal variation and external interferences. In this case, variation is treated as epistemically uninteresting noise. More recent research, however, attaches an explanatory significance to the fact that molecular mechanisms operate at scales at which thermal noise is as potent as mechanical, electrostatic, and chemical interactions. On this account, biological mechanisms are not functioning despite noise, but rather because of noise. There is also evidence suggesting that biological mechanisms have evolved to both suppress and increase molecular noise. Some mechanisms are well insulated, highly modular, and less noisy, while other

---

[7] McAllister (1997) argues that statistics fails to provide objective means of defining what counts as noise. He points out that more than one regression model may pass a statistical significance test and that thresholds of significance defining what counts as noise can be arbitrarily varied such that infinitely many 'signals' can be extracted from any given data set.

mechanisms overlap with one another, are considerably noisier, and generate more complex stochastic outcomes. These two interpretations make different testable predictions, which in principle can be disentangled by empirical investigation.[8]

2.5 The Data-Phenomena Distinction

I opened the section with a remark about phenomena being commonly thought of as both constitutive of empirical reality and objects of explanation. Data satisfy this description. They correspond to that segment of empirical reality supporting inferences about underlying causal structures, being epistemically significant in a way mere observations are not. This is not a coincidence. Empirical research is regulated by methodological principles explicitly aimed at ensuring that data is informative about the causal structure of the world. Standardized measurement techniques, detailed descriptions of experimental setups, and attempts to improve reproducibility are all meant to ensure that data fulfill their epistemic purpose. Besides, consistently reproduced data account for many phenomena indisputably recognized as such in science. Inflammatory responses, immunization, hemagglutination, and the antiinflammatory effects of drugs are just a few examples from immunology. More generally, reproducible data are both something that cannot be dismissed as mere coincidence, and therefore demand an explanation, as well as something in principle explainable, since their association with a replicable experimental setup offers a promising starting point for further investigation.

Despite its prima facie plausibility, the proposal that phenomena are data goes against a relatively firm philosophical consensus claiming the opposite. This consensus is motivated in part by the tacit assumption that data amount to what logical positivists call 'observations', namely end products of measurements, such as perceptual reports and measured variable values. I think this assumption is mistaken. If the term 'data' refers to the kind of empirical results typically reported in scientific papers, then data are not mere observations. By themselves, the latter don't have much scientific significance. Relevant empirical information, from which something can be inferred about the causal structure of the world, is systematically structured as contrastive data sets, where each individual data point is not just the measured value of a variable but has attached to it labels specifying the method of measurement and the experimental setup in which the measurement was conducted. Reproducible data imply an even

---

[8] For an overview of the two approaches, see (Baetu 2017a). An edited volume by Kupiec et al. (2011) covers philosophical perspectives on the explanatory role of noise and stochasticity in biology. For a review of the scientific literature on molecular noise, see (Rao et al. 2002).

more complex structure involving networks of correlations and causal dependencies between the values of the variables targeted by measurements and interventions, and the values of label-variables on the basis of which an experimental setup is said to have been replicated and a method of measurement/intervention standardized.

The view that data is inseparable from experimental setups is reminiscent of Ian Hacking's (1983, 226) claim that a phenomenon "does not exist outside of certain kinds of apparatus," constituting what one may justifiably call "technology, reliable and routinely produced." The view I advocate does, however, diverge from Hacking's in two respects. First, Hacking argues that phenomena exist only under controlled laboratory conditions, which are seldom approximated in nature. Yet what happens in nature must also be measured and measurements always require some form of experimental control. Thus, what Hacking takes to be a categorical dichotomy between data and phenomena is more a question of degree of experimental control doubled by a potential problem of extrapolation, namely the worry that knowledge produced by studying well controlled experimental setups may fail to reflect what happens in less thoroughly controlled experimental setups. A second difference concerns the degree of reproducibility. Hacking's focus on reliability and tightly controlled experimental setups captures the intuition that only data highly informative of the causal structure of the world count as phenomena. There are, however, good reasons to resist imposing too stringent standards of reproducibility. By denying variable data the status of phenomena, one incurs the risk of dismissing the investigation of experimental setups encompassing inherently noisy causal structures such as heterogeneous populations, complex systems, and stochastic mechanisms. Also, many everyday situations can be construed as poorly controlled experimental setups; dismissing the study of such setups curtails the potential for practical applications.[9]

Bogen and Woodward (1988, 317) draw a distinction between data and phenomena on the grounds that data are "idiosyncratic to particular experimental contexts, and typically cannot occur outside of those contexts," while phenomena "have stable, repeatable characteristics which will be detectable by means of a variety of different procedures, which may yield quite different kinds of data." For example, since a variety of inflammatory responses are

---

[9] I have argued elsewhere that the elucidation of mechanisms by experimental methods requires only that data be reproduced often enough relative to the lifetime of a research program (Baetu 2013). How often is 'often enough' depends on how long it takes us to elucidate the causal structures underpinning data by experimental methods. Given enough time, even rare phenomena that we ordinarily take to be chance occurrences are in principle accessible to empirical investigation.

commonly grouped under the category of 'inflammation', one may insist that the phenomenon studied by immunologists is a general phenomenon of inflammation, which may be investigated in a cell or an animal model, using ultraviolet radiation, a virus, or some other harmful stimulus as an inducer, and by assessing symptoms via a multitude of methods. Building on such considerations, Bogen and Woodward conclude that scientific knowledge is a three-layer structure, with a theory-*explanans* at the top, a phenomenon-*explanandum* in the middle, and, at the bottom, data construed as evidence for the *explanandum*.

I think the notion that phenomena are generalizable across multiple experimental setups is only part of the story. In later stages of research, phenomena are both lumped and split into various categories as knowledge of their mechanistic underpinning is unveiled (Craver 2007, 123–24; Craver and Darden 2013, 60–62). For example, symptom-based characterizations of diseases are systematically revised in light of subsequently acquired knowledge of physiopathological mechanisms, leading to frequent splitting and lumping of preexisting categories (splitting of diabetes into several types, lumping of Asperger's with other conditions, etc.). In early stages too, the mere fact that experimental setups are modified in order to increase reproducibility prompts some modest recharacterizations of phenomena.

Nevertheless, even if phenomena are recharacterized, the conclusion that phenomena have nothing to do with measurements and experimental setups, but are instead the causal structures these measurements probe, is problematic. For one thing, the mechanistic literature is unanimous in insisting that mechanisms are individuated according to the phenomena for which they are responsible. But Bogen and Woodward tell us that phenomena are individuated in terms of 'repeatable characteristics detectable by many different procedures'. Since these characteristics must be the causal structures probed by measurement procedures, phenomena are individuated in terms of underlying causal structures; that is, in terms of mechanisms. There is a way out of this impasse if we consider the fact that Bogen and Woodward adopt a retrospective, 'end of the scientific inquiry' standpoint assuming that a significant body of knowledge about the causal structure of the world is already available. Yes, arrived at this point, many phenomena have already undergone more or less drastic recharacterizations reflecting well established knowledge of their mechanistic underpinnings; however, we must also account for the prospective standpoint of the experimental scientist. Set alongside a minority of phenomena predicted on theoretical grounds, phenomena have an empirical origin that long precedes their eventual recharacterizations in light of accepted explanations. Bogen and Woodward don't tell us anything about these phenomena, which are, after all, what motivates most scientific inquiry in the first place.

Inasmuch as researchers typically characterize *explananda* in the absence of any putative *explanans*, I think it is correct to assume that most phenomena begin their epistemic career as data reproduced when experiments are replicated. As knowledge accumulates, data is sometimes grouped in radically novel ways, for instance, lumped together given evidence of a common mechanism, or split into distinct phenomena if underpinned by different mechanisms. An explanatory-driven classification, however, does not annihilate the individual *explananda* items it classifies. Rather, in the case of lumping, it revises the expected domain of validity of certain extrapolations (i.e., how far findings can be generalized), while in the case of splitting it gives an impetus for designing new, more precise experimental setups. For instance, the lumping together of various inflammatory response phenomena under the heading 'inflammation' reflects the fact that all these responses involve similar symptoms and are mediated in part by the same mechanisms. In practical terms, researchers expect that core mechanistic components are shared by all inflammation responses, which in turns fuels the working assumption that universal treatments of undesired inflammatory responses can be developed by targeting these components. This does not mean, however, that the phenomenon of ultraviolet-induced erythema (sunburns) is identical with that of septic shock or peanut allergy, or that one and the same mechanism is responsible for all these phenomena. There are important differences, both symptomatic and mechanistic, which dictate that the elucidation of each phenomenon requires a distinct experimental setup (Netea et al. 2017).

A second concern is that by shifting the *explanandum* from measurements to measured causal structures, as Bogen and Woodward do, one inevitably also shifts the locus of explanation from the identification of causal structures to a theoretical account of these causal structures. This is unrepresentative of explanations in biology. A typical biological *explanandum* is a reproducible data set, for instance, the repeated reduction of erythema in human or animal subjects taking ibuprofen, as compared to control subjects. The *explanans* is a causal structure, in this case the mechanism of action of ibuprofen, which has something to do with ibuprofen interfering with the biological mechanisms underlying inflammatory responses. There is no further theoretical *explanans*, no overarching theory of mechanisms of action of ibuprofen or those of inflammation. From a mechanistic perspective, the phenomenon to be explained is not the measured causal structure, but rather the data produced by measurement experiments.[10]

---

[10] I think it is correct to assume that there are no true theories in biology, at least not of the sort found in physics, but only broad theoretical constraints dictating, for example, that the mechanisms of action of pharmacological agents must have a molecular component. Since such

Perhaps the most important reason for distinguishing data and phenomena is the worry that data noisiness hampers systematic explanation. James Brown (1994, 125) characterizes phenomena as "abstract entities which are (or at least correspond to) visualizable natural kinds." In addition to the reference to 'natural kinds', which suggests that phenomena correspond to an ontologically distinct category, abstraction refers to a data processing method in virtue of which an explainable 'signal' is identified amidst epistemically uninteresting noise. This characterization builds on Bogen and Woodward's (1988) influential argument that measurements never succeed in conveying specific information about the feature of reality targeted by a measurement, but invariably also report on a variety of idiosyncratic causal interferences at work on the occasion of a particular measurement. Hence, data never fall under the scope of any single explanation but require a combination of diverse explanations varying from case to case.

Data variability is indeed a major source of concern in experimental practice. This is not, however, because of frustrated attempts at systematic explanation. Systematic explanation is not the norm in biology. The fact that a mechanism produces a certain phenomenon in one biological system offers no firm guarantees that similar mechanisms produce similar phenomena in other biological systems (Baetu 2016a; Bechtel 2009; Tabery 2009). Even in the same biological system, the same phenomenon may be simultaneously generated by more than one mechanism, with some mechanisms having a higher biological relevance in one situation, but not another. Subtle and not so subtle differences are omnipresent, from variations in the highly conserved mechanisms responsible for biological activities shared by all living things (e.g., slight variations of the genetic code, or the more significant differences between prokaryotic and eukaryotic genome expression mechanisms making possible the use of antibiotics), to differences in the mechanisms underpinning similar biological functions (e.g., acquired immunity is a shared characteristic of all jawed vertebrates, yet the mechanisms underpinning it vary in different species), to differences between individuals of the same species (AIDS pathogenesis in

---

constraints are insufficient to predict phenomena and mechanisms with any satisfactory degree of detail, the discovery of biological phenomena and their mechanisms relies heavily on experimental methods. This is not to say that biological mechanisms cannot or should not be investigated from a theoretical perspective. My point is that an account of phenomena should not obscure the fact that the identification and description of mechanisms is an important type of explanation in biology. Once mechanisms are hypothesized or elucidated, theoretical considerations may play a role in predicting certain properties of these mechanisms. Possible examples include hysteresis, predicted by quantitative-dynamic models of molecular pathways and mechanisms, stochastic phenotypic variation in clonal cell populations, predicted by the impact of thermal noise on biochemical processes, and differences in the rates of certain biochemical reactions predicted by free-diffusion and macromolecular crowding models.

humans is variable due to resistance mediated by truncated cell receptors, the presence of specific mechanisms of defence, etc.).

Rather, the concern for data variability stems from the fact that uncontrolled variation undermines the validity of contrastive inferences. As argued in Section 2.3, this concern is addressed by attempts to increase experimental control in order to enhance reproducibility and thus render data more informative of their *explanans*. A more general response to the worry that data are unamenable to systematic explanation is that, in the end, data always needs to be accounted for in all its gory details. A 'signal', or systematically explainable component of data, can be abstracted only in as much as the remaining noise is also accounted for. Echoing the classical theory of errors, Bogen and Woodward argue that variations in measured values of the melting point of lead are attributable to a multitude of weak, nondirectional interferences muddling the main causal signal. Since these many and diverse interfering causes are not, and most likely can never be specified, they conclude that data are, by and large, unexplained and unexplainable. This is not exactly true. Even though the exact nature of causal interferences is not specified, a statistical model embodies substantive claims about how data is generated, thus offering a putative explanation of why data scatter in a particular way (Burnham and Anderson 2002). This explanation is supported not only by its consistency with the actual variation of measured values, but also by the fact that measurement precision can be increased by improving measurement techniques and experimental setups. In other words, modeling noise as measurement error is a working hypothesis making a claim about physical reality. If this hypothesis turns out to be false, the whole explanatory schema needs to be revised, starting with the signal-noise distinction. For instance, measurements of DNA denaturation ('melting') temperature also vary, yet unlike variations in lead melting temperature, the former are not explained as measurement error alone, but are also attributed to the fact that there is no such thing as a unique true value of the measured variable. Denaturation temperature is determined by the number of hydrogen bonds to be broken, which in turn depends on the length and nucleotide composition of each DNA molecule.

2.6 A Working Characterization of Phenomena

What the above attempts to make sense of the notion of 'phenomenon' suggest is that, despite a much-argued-for distinction between data and phenomena, the two are intimately linked to a much greater extent than most philosophers are willing to admit. The fact that phenomena are described as less noisy, reliable, reproducible, and recurrent *explananda* does not preclude them from being data. Conversely, noisiness does not automatically disqualify data from ful-filling the epistemic role of *explananda* typically attributed to phenomena. None of these descriptors apply exclusively to phenomena or data. The situation is further complicated by the fact that phenomena are a dynamic category. In the early stages of research, any data generated by a controlled experiment is a potential phenomenon. In the most fruitful phase of experimental research, phenomena typically amount to data reproduced across one or, more commonly, a series of experimental setups, although experimental research neither excludes the investigation of irreproducible and noisy data nor is incompatible with the possibility that such data may be phenomena amenable to systematic explanation. In later stages, the identification of causal factors and the eventual elucidation of mechanisms prompts more-or-less dramatic recharacterizations of phenomena. Finally, a small number of phenomena are predicted top-down, from theories and explanatory accounts already enjoying substantial empirical support.

Since a characterization of phenomena is necessary for a philosophical dis-cussion of mechanisms, I propose adopting an approximation that will do the job in the case of biology. The focus of this Element is primarily on the period of sustained experimental research that best describes contemporary biology, especially cell and molecular biology, as well as biomedical and clinical research. Thus, if I were to distill an element of commonality among phenom-ena characterized at this stage of scientific inquiry, I cannot be too far off in claiming that phenomena are data reproduced when experiments are replicated. According to this view, data and phenomena are indistinguishable from an epistemic point of view, both fulfilling the same functional role of *explanandum* in the overall economy of scientific knowledge. From a metaphysical stand-point, both data and phenomena are nothing else but measurements, albeit ones conducted in such a way as to serve a specific epistemic purpose, namely that of revealing something about the causal structure of the world.

3 How Do Mechanisms and Phenomena Relate to One Another?

A mechanism is said to be responsible for a phenomenon in the sense that it produces or underlies that phenomenon. Two metaphysical accounts are stan-dardly offered. An etiological account captures the intuition that mechanisms should be causally relevant to phenomena given that the discovery of mechan-isms relies on experiments demonstrating causal relevance. The implication, however, that phenomena are end-point outcomes of causal chains is at odds with experimental characterizations of phenomena. Alternatively, a constitutive account captures the notion that phenomena are behaviors of systems explained by referring to the behavior of their parts. Yet this account raises a profound puzzle about how evidence for causal relevance can justify a metaphysical interpretation postulating noncausal relationships between mechanisms and phenomena. I reject both accounts in favor of a third one, stating that a phenomenon is nothing else but the data generated by measuring a mechanism. Conversely, a mechanism is a causal structure linking the variables probed by the measurements involved in the description of a phenomenon in relationships of correlation and causal dependency.

3.1 Causal Relevance

In biological sciences, the elucidation of mechanisms relies heavily on experimental interventions meant to demonstrate relationships of causal relevance among variables or, if a physical interpretation is available, the causal relevance of various factors to an outcome of interest (Craver 2007). Following up on the example of inflammatory responses, the discovery of prostaglandins in edema fluids (inflammatory exudates) led researchers to hypothesize that these hormone-like compounds are components of the biological mechanisms of inflammatory responses. This hypothesis was supported by experiments showing that, indeed, when injected intradermally, prostaglandins cause an inflammatory response. This result prompted further speculation that aspirin, ibuprofen, and other nonsteroidal antiinflammatory drugs block the production of prostaglandins. Again, this proved to be correct. John Vane famously showed that aspirin blocks prostaglandin synthesis in guinea pig lung homogenates.[11]

In the above examples, causal relevance was established by comparing outcomes in two situations, a test condition in which a factor under investigation is present and a control in which it is not. In order to conclusively demonstrate that the factor is causally responsible for the difference in outcome, alternative explanations had to be ruled out:

(1) The possibility that factor and outcome are divergent effects of a common causal background or changes in outcome cause changes in the factor was ruled out by conducting an intervention on the factor while the outcome is monitored for changes. If factor and outcome are divergent effects or the latter is an effect of the former, then the intervention on the factor is not expected to have an impact on the outcome. However, if there is a causal pathway linking factor and outcome as upstream cause to downstream

---

[11] For a historical overview, see (Botting 2010).

effect, then interventions on the factor are expected to result in changes in the outcome.

(2) The possibility that some other difference between the two conditions (a confounder) is responsible for the difference in outcome was ruled out by ensuring that test and control conditions are comparable in all relevant respects minus the factor manipulated in the experiment. In basic science, comparability is achieved by systematically removing differences between biological systems in order to generate genetically and phenotypically homogeneous organism strains and cell-line clones. Standardization and operationalization are meant to ensure that the same experimental setup is replicated in the two arms of the experiment, while a parallel design in which test and control are simultaneously deployed side by side further minimizes unequal exposure to external sources of interference. The goal of these strategies is to develop an experimental model of a phenomenon, which is expected to capture a class of comparable objects relative to which the outcomes of interventions can be interpreted in terms of causal relevance.[12]

(3) Finally, correct attribution of causal efficacy requires an accurate intervention targeting only the factor under investigation and no other factors that may contribute to differences in outcome. This is often demonstrated by including additional positive and negative controls. For example, placebo interventions in the control arm of the experiment are meant to ensure that the relevant difference maker is not some generic lab procedure, such as gently shaking cells, but rather the investigated factor, say, a virus, which is added while gently shaking the cell suspension.

The above desiderata are captured by interventionist accounts of causation, which state that evidence for causation requires that an intervention on causal factor X must change the outcome Y without changing any other variable that is a cause of Y – that is, without directly changing Y, or any other variable along the causal pathway from X to Y, or by simultaneously intervening on convergent causal pathways leading to Y (Woodward 2003, 94–99). Similar desiderata are

---

[12] In clinical research, causal comparability cannot be assumed. In randomized controlled trials, causation is inferred by estimating the probability of generating differences in outcome in virtue of random allocation. By itself, statistical inference only yields a statement about the probable existence of a cause other than random variation, without revealing the identity of this additional cause. Inasmuch as it can be demonstrated, however, that the experiment did not introduce any differences other than random variation, this cause can be identified as the treatment administered to the test group. This requirement is satisfied by ensuring that the allocation of patients to test and control groups (the experimental intervention) is not biased, which is achieved by randomizing the study.

emphasized in the experimental and clinical methodology literature (Hill 1965; Leighton 2010).

3.2 The Etiological Account

The fact that mechanisms are elucidated by means of experiments designed to demonstrate causal relevance speaks in favor of a causal relationship between mechanisms and phenomena. It is not surprising, therefore, that most authors agree that at least some mechanisms amount to causal chains terminating in the phenomena to be explained. An example cited in the literature is the excessive number of trinucleotide (CAG) repeats in the huntingtin gene, which is viewed as part of the etiological mechanism causing Huntington's disease (Craver 2007, 107–08). The phenomenon is construed as the presence of a set of symptoms, namely chorea accompanied by cognitive impairments, depression, and compulsive behavior. The mechanism of Huntington's disease is viewed as the causal pathway initiated by an excess number of repeats and terminating with the manifestation of symptoms.

Despite its initial appeal, the etiological account faces a serious difficulty. The mere description of a time-point outcome, such as the occurrence of a set of symptoms, doesn't provide much information about its causes. The causal pathway leading to the outcome and the starting point of the pathway are left unspecified. The Big Bang is a cause of Huntington's symptoms to the same extent as mutations in the huntingtin gene. Moreover, Huntington's disease belongs to a large family of choreic syndromes, all characterized by very similar symptoms, but associated with different gene mutations (e.g., Huntington's disease-like syndromes) or even nongenetic risk factors, such as stroke or brain lesions caused by autoantibodies (Sydenham's chorea).

In the absence of additional constraints, it is not clear how the mechanism responsible for such a phenomenon can be individuated and its spatiotemporal boundaries delineated. This raises the suspicion that the description of the phenomenon is incomplete. If outcomes such as the symptoms of Huntington's came to be identified as a phenomenon as opposed to being treated as a handful of unfortunate happenings, it must have been in virtue of the reproducibility of the symptoms in the context of a replicable experimental setup. It is not difficult to guess that even though the incidence of Huntington's in the general population is extremely low, and for this reason the disease remained unknown to the medical community until the nineteenth century, those affected must have realized that the symptoms run in the family. It is indeed this fact that caught the attention of George Huntington, who discovered this rare disease by reconstructing the pedigree of an affected family, outlining what eventually came to be recognized as an autosomal-dominant pattern of inheritance.

A crucially important point is that Huntington didn't just document out-comes. He defined a replicable experimental setup, consisting of parent-offspring lineages, in the context of which a pattern of inheritance, as described by the incidence of outcomes in the descendants of affected parents, is consistently reproduced. The phenomenon investigated ever since Huntington first characterized the disease is therefore not the occurrence of a set of symptoms, but rather the occurrence of a set of symptoms in individuals having a certain kind of family history, namely a history of autosomal-dominant chorea. A comparison of genetic markers between healthy and affected individuals with the relevant family history revealed that symptoms are associated with mutations on chromosome 4, which were eventually narrowed down to a smaller segment containing the huntingtin gene. A comparison between huntingtin gene sequences from patients belonging to families with a history of autosomal-dominant chorea revealed a correlation between the number of CAG repeats and the ages of onset and death. These findings, in conjunction with results showing that the severity of symptoms correlates with brain atrophy and that mutated huntingtin protein is neurotoxic in cell and animal models, strongly support a genetic etiology.[13] Note therefore that even though CAG repeats are casually referred to as 'the cause' of Huntington's disease, they are in fact nothing else but a causally relevant factor. Moreover, this factor doesn't precede the Huntington's disease phenomenon, but acts within its temporal duration. As stated earlier, the phenomenon under investigation is the occur-rence of chorea in a population of patients with a specified type of family history. The mutated huntingtin gene doesn't precede the family lineages in which the symptoms are documented.

In more general terms, the notion that phenomena are time-point outcomes is incompatible with the experimental methodology underpinning studies aiming to determine etiology. From an experimental point of view, outcomes are nothing else but measured values of variables at particular points in time. No experimental setup is specified. We don't know if measurements, say of Huntington's symptoms, are to be conducted in the same individuals, in indivi-duals taken from the general population or from a more specific subpopulation, at the age of onset, death, or some other point in time, and so on. The only thing we are told is that an outcome is sometimes documented and that this is the phenomenon to be explained. The difficulty here is that, by themselves, out-comes are not data informative of the causal structure of the world. In order to

---

[13] A historical review of the discovery of the genetic basis of Huntington's is covered in (Bates 2005). Given the limited potential for experimental interventions in humans, evidence for causation is indirect, relying on extrapolations from surrogate biological systems (Baetu 2016a; Germain and Baetu 2017; Steel 2007).

inform on causal determinants, outcomes must be reproducible. In turn, reproducibility relies on the specification of an experimental setup replicating the conditions in which measurements are conducted. As a result, one can only consistently document and ultimately elucidate the causes of phenomena such as autosomal-dominant inherited chorea or CAG-extension induced neuronal death, while mere outcomes such as chorea or neuronal death are subject to extreme variation depending on when, where, and under which conditions these outcomes happen to be measured.

3.3 The Constitutive Account

In the mechanistic literature, phenomena are seldom construed as outcomes. They are usually depicted as regularities, patterns, or behaviors associated with biological systems (Bechtel and Richardson 2010; Craver 2007; Machamer et al. 2000). The characterization proposed in Section 2, stating that phenomena are data reproduced when experimental setups are replicated, is compatible with such depictions since it entails consistent correlations among variables. Some mechanists further emphasize that phenomena have intrinsic spatiotemporal and dynamic features (Bechtel and Abrahamsen 2010). These features too can be linked to the specification of a replicable experimental setup. Failure to include spatiotemporal data labels severely undermines reproducibility. For instance, similar degrees of erythema can be reproduced only if the interval of time between induction and assessment is kept constant. Finally, the fact that data change when experimental setups are altered supports the notion that at least some phenomena are recurrent causal dependencies between variables (Woodward 2002). For example, given that alterations involving interventions on variables describing an experimental setup (e.g., the duration of ultraviolet exposure) are a common strategy for enhancing reproducibility (consistently generate inflammatory responses of similar intensity), experimental models optimized for reproducibility are typically expected to document causal dependencies.

These characteristics of phenomena indicate that mechanistic explanations are not aimed at identifying antecedent causes, but rather at elucidating the mechanisms 'underlying' input–output behaviors. According to an influential proposal, this underlying relationship is best understood in terms of constitution. For example, Carl Craver and William Bechtel (2007) argue that the phenomenon of light transduction in the eye consists of, as opposed to being caused by, a hierarchical structure of mechanisms involving parts behaving in certain ways, much in the same way the temperature of a gas consists of, and is not caused by, the mean kinetic energy of gas molecules.

The immediate difficulty is that it is not clear how constitution relationships can be inferred given that experiments designed according to the methodology described in Section 3.1 can only provide evidence for causal relevance. Craver proposes an ingenious way of bridging the gap between the two:

> a component is relevant to the behavior of a mechanism as a whole when . . . [the] two are related as part to whole and they are mutually manipulable . . .: (i) X is part of S; (ii) in the conditions relevant to the request for explanation there is some change to X's ϕ-ing that changes S's ψ-ing; and (iii) in the conditions relevant to the request for explanation there is some change to S's ψ-ing that changes X's ϕ-ing. (2007, 153)

For instance, the whole S could be a guinea pig organism; S's behavior ψ, an inflammatory response; part X, prostaglandins; and X's behavior ϕ, the binding of prostaglandin receptors, which triggers a signaling cascade leading to the expression of several gene products involved in inflammatory responses. The fact that prostaglandin injections result in inflammatory responses may be construed as a bottom-up intervention on a part having an effect on the whole, while ultraviolet exposure could be a top-down intervention on the guinea pig-whole affecting its prostaglandin-parts.

The mutual manipulability account presupposes a system or whole exhibiting a specific behavior, along with a decomposition strategy yielding parts exhibiting their own specific behaviors. Something needs to be said about how these items are defined. In the examples discussed in the literature, a system is typically an organism, a part of an organism (cell, organ), or a population of organisms. This conception, relying on tacit intuitions about what counts as a biological individual, is rather restrictive, since it entails that mechanisms cannot extend beyond the spatiotemporal boundaries of a given biological system. For example, Earth's magnetic field cannot be part of the biological mechanism of bird navigation. As a result, a somewhat arbitrary distinction is drawn between triggering conditions located outside a biological system and relegated to various etiological mechanisms, and the constitutive mechanism, which is strictly confined within the boundaries of the system. Behaviors, on the other hand, are understood quite liberally to include stimulus–response causal sequences, input–output correlations, clusters of symptoms, and networks of causal and correlated factors. Although a behavior is tacitly assumed to be the behavior of a biological system, the implication here being that one first chooses a system and then describes its behavior, an equally legitimate approach is to first choose a behavior and then define the system as a collection of variables or factors relevant to that behavior. For instance, two associated factors often constitute a system worthy of further investigation. The main advantage of this approach is that the boundaries of wholes and mechanisms are determined a posteriori, thus avoiding the need to rely on prior intuitions. Moreover, these boundaries can be revised as new causal and correlated factors are discovered and the need to distinguish between triggering conditions and constitutive mechanisms vanishes. On the negative side, nothing guarantees that systems thus defined include all the mechanistic components necessary to that behavior. For instance, a system defined based on the association of factors may exclude from its description the common cause responsible for their association.

The characterization of phenomena developed in Section 2 offers a practicable middle way between these two extremes. The whole under investigation is an experimental setup including, along with a biological system, parts of the environment relevant to a biological phenomenon. For example, a cell model of biological activity or disease includes not only cells extracted from an organism, but also an artificial growth medium mimicking key aspects of the physiological conditions characteristic of the living organism. Likewise, an animal model, for instance, of inflammatory responses, includes elements of the environment, such as ultraviolet radiation, allergens, or pathogens, in addition to the animal organism (Figure 4). The initial choice of an experimental setup may very well be motivated by tacit intuitions about how reality is structured into objects, events, and processes. Nevertheless, an empirical characterization emerges after subsequent attempts to modify the experimental setup in order to enhance the reproducibility of a phenomenon.

Something also needs to be said about how mechanism and system relate to one another. Craver and Bechtel do not always draw a clear distinction between the two, which may be taken to imply that the system is the mechanism. On second thought, however, this cannot be right. Not all the parts of the system are relevant to its behavior. For one thing, not all decomposition strategies yield the right kind of parts, and even if the right decomposition strategy is adopted, some parts of the system may play a role in respect to different behaviors of the system or have no role whatsoever. For instance, a whole guinea pig is not required to produce prostaglandins when exposed to UV radiation; even if it loses the tip of its left ear, the guinea pig (and, if artificially kept alive, the bit of ear too) can still develop an inflammatory response. Presumably, this is where Craver's notion of constitutive relevance becomes important: only those parts of the system standing in a relationship of mutual causal relevance to the behavior under investigation are components of the mechanism responsible for that behavior. It may therefore be concluded that a mechanism includes only a subset of the parts of a system, namely those involved in a mutual manipulability relationship with the system (Kaiser and Krickel 2017). This conclusion is similar to my own claim that an

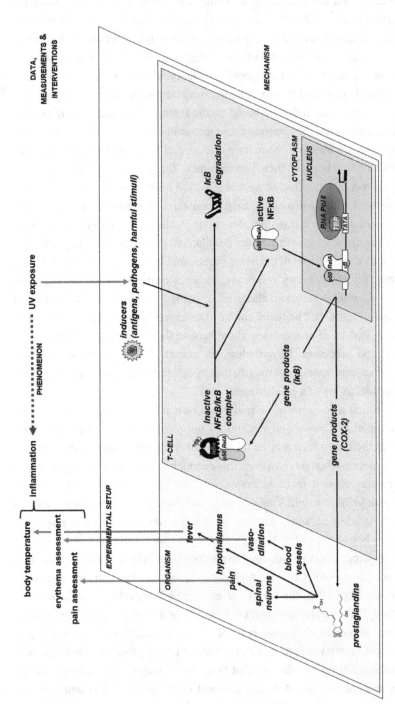

Figure 4 Experimental setups (systems), phenomena (behaviors of systems), and mechanisms

experimental model contains within its spatiotemporal boundaries the causal structures responsible for a phenomenon.

Since mutual manipulability relies on causal relevance tests, it seems natural to conclude that the relationship between mechanism and phenomenon must be one of reciprocal causal dependency (Leuridan 2012). Nevertheless, Craver and Bechtel insist that the requirement for part–whole relationships has unpalatable consequences for a causal interpretation, chief among which is the fact that cause and effect are no longer distinct events (Craver 2007, 153–54; Craver and Bechtel 2007, 552–54). Yet if it is indeed the case that two variables stand in part–whole relationships, then it cannot be conclusively demonstrated that interventions on a variable have an effect on the other variable. In other words, it is impossible to demonstrate mutual manipulability in the first place. If lower levels supervene on higher ones, as implied by the part–whole constituency requirement, then top-down interventions invariably have an effect on the behaviors of both wholes and their parts (Baumgartner and Gebharter 2016; Romero 2015). Since several variables are targeted at the same time, the accuracy of the intervention is compromised (Section 3.1, condition 3) and it cannot be concluded that the effects observed in the dependent variable are indeed due to experimental interventions on the independent variable and not to direct interventions on the dependent variable. This argument further highlights the fact that talk about interventions on and measurement of parts and wholes is remarkably vague. If I push a cup of coffee over the edge of the desk, do I intervene on the cup? Or do I intervene on the room, Earth, the solar system, the whole universe, and all those many other things of which the cup is a part? Or perhaps I intervene on the parts of the cup, all the way down to the elementary particles of which the cup is made? And where do I measure the effect? The cup is certainly affected, and so am I, along with the rest of the universe, which must come to terms with the realization that the cup is now shattered on the floor, and all the parts and particles of which the cup is made, as they too are now all over the floor.

One culprit behind these ambiguities is a failure to correctly identify the independent and dependent variables targeted in an experiment. Craver (2007, 166–70) describes the training of rats in a Morris maze as a top-down experiment in which the behavior of rat-wholes is manipulated (the independent variable, corresponding to the factor tested for causal relevance) and effects on the behavior of parts, such as the long-term potentiation of certain synapses, are measured (the dependent variable). Yet this description is incorrect. What varies between test and control conditions (the independent variable) is not the rats or their behaviors, which are assumed to be comparable at onset between test and control conditions (Section 3.1, condition 2), but rather the maze, more specifically the variable 'place of escape platform', which takes the values 'constant' or 'random'.[14] The same incongruency is present in the inflammatory response example. What varies between test and control conditions is not the guinea pigs or erythema, but the intensity of ultraviolet radiation, which is the input condition of the phenomenon of ultraviolet-induced erythema.

Craver seems to be aware of this caveat when he argues that "[o]ne intervenes on S's ψ-ing by intervening to provide the conditions under which S regularly ψs. Top-down experiments intervene in this way" (2007, 146). Unfortunately, this proposal suffers from the same methodological incongruity plaguing the etiological account. The phenomenon under investigation is a behavior amounting to at least an association of variables. Huntington's disease is not the mere occurrence of a set of symptoms, but a consistently reproducible set of symptoms in a family lineage; an inflammatory response is not an occurrence of symptoms, but a consistently reproducible set of symptoms in response to certain stimuli; spatial memory too is not measured solely as the time it takes a rat to navigate a maze, but the time it takes a rat to navigate a maze given prior exposure to environmental clues. In contrast, the experiments involved in the elucidation of the mechanisms target individual variables, such as ultraviolet exposure, erythema, or the position of the escape platform in a water maze. It would seem therefore that talk about bottom-up and top-down interventions plays on an ambiguity whereby the manipulation or measurement of a variable involved in the description of a phenomenon is arbitrarily equated with the manipulation or measurement of a phenomenon.

3.4 The Causal Mediation Account

I suspect that the inconsistencies plaguing the constitutive account stem from a prior commitment to the view that phenomena are macro-level states consisting of micro-level mechanisms. What exactly constitution amounts to varies from author to author, and could be understood as 'supervene on', 'are realized by', 'are identical with', or 'are made of'. Whatever it is, constitution is definitively not causal. In contrast, experimental methodology is explicitly designed to generate evidence for causation. This creates a tension between evidence and metaphysics, which, as shown in the previous section, is not resolved in a satisfactory manner.

---

[14] Morris (1981, 242) mentions that a rat was excluded from his well-known experiment on spatial memory because it was found to be different from the other rats in a respect relevant to the measured outcomes 'escape latency' and 'directionality of tracks': "One rat was found to have difficulty in swimming during the first Pretraining session and was replaced with another animal."

There is an obvious solution to this problem. Instead of first committing to a noncausal interpretation and then attempting the impossible task of demonstrating that evidence for causation can somehow be used to demonstrate something else than causation, one can simply let go of ready-made metaphysical intuitions borrowed from physics and the philosophy of mind and consider the relationship between mechanisms and phenomena from the strictly experimental standpoint adopted throughout Section 2.

In the inflammatory response example, the phenomenon to be explained is ultraviolet-induced erythema, that is, the consistently reproducible induction of erythema in response to exposure to ultraviolet radiation. What is construed under the mutual manipulability account as a 'bottom-up' intervention can just as well be described as the phenomenon of prostaglandin-induced erythema, namely the consistently reproducible induction of erythema in response to a local increase in the concentration of prostaglandins. The 'top-down' intervention can likewise be described as the phenomenon of ultraviolet-induced increase of local prostaglandin levels. Instead of asking whether one phenomenon supervenes, is realized by, is identical with, or made of another phenomenon, we can simply ask whether the causal pathways involved in ultraviolet-induced prostaglandin synthesis and prostaglandin-induced erythema are constitutive, or parts of the causal pathway linking ultraviolet exposure to erythema. This question can easily be reframed in terms of causal mediation (Baetu 2012a; Harinen 2014). To ask whether prostaglandin synthesis is constitutively relevant to the phenomenon of ultraviolet-induced erythema is to ask whether prostaglandin synthesis is a causal intermediary along a causal pathway linking the ultraviolet exposure-input to the erythema-output.

The advantage of asking questions about causal mediation is that there is a well-established experimental methodology for answering them. In contrast, if we insist on asking questions about supervenience, realization, identity, or composition, then we must accept the fact that experimental research doesn't have any answers to offer. In other words, there is a straightforward way of empirically testing our metaphysical hunches about causal relationships, but not those about noncausal constitutive relationships. How then is causal mediation demonstrated? The general strategy is to conduct a knockout-type experiment whereby two factors, usually the initial conditions and a putative mechanistic component, are simultaneously manipulated and the effects on a third variable, usually the output conditions, are observed (Figure 5). One such experiment demonstrated that when prostaglandin synthesis is blocked, ultraviolet exposure fails to cause erythema (Langenbach et al. 1999). The fact that a failure to synthesize prostaglandins disrupts the association between ultraviolet exposure

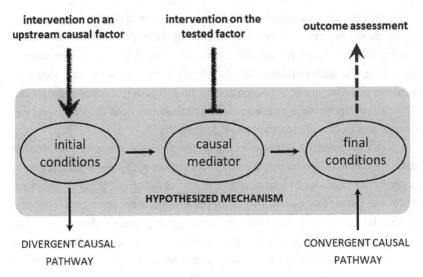

Figure 5 Sorting of factors by a two-variable knockout experiment

and erythema demonstrates that prostaglandins are causal intermediaries mediating ultraviolet-induced erythema.

According to the causal mediation account, the mechanism responsible for the phenomenon of ultraviolet-induced erythema is a causal structure linking input (ultraviolet exposure) and output (erythema) conditions, 'linking' being understood here as 'not allowed to vary from one iteration of the experiment to the next'. To be part of, or constitutively relevant to the mechanism is to be a causal intermediary along this causal structure. In this example, the fact that prostaglandins are causal intermediaries mediating ultraviolet-induced erythema was taken as evidence that prostaglandins are components of the mechanism responsible for the phenomenon of ultraviolet-induced erythema.

One distinctive feature of this account is that both the 'higher-level' description of a phenomenon and that of its 'underlying' mechanism refer to the same causal structure. The difference lies in the fact that a phenomenon is described by means of interventions and measurements probing very few aspects of this structure, while intermediary causal stages remain unknown. For this reason, descriptions of phenomena are sometimes assimilated to 'black boxes' (Craver 2007; Craver and Darden 2013; Darden 2006b; Machamer et al. 2000). A mechanistic description, on the other hand, is based on additional interventions and measurements targeting variables or factors characteristic of the intermediary stages of the mechanism. By disrupting the inner workings of the mechanism in ways that affect downstream variables used to test for causal relevance, causal factors initially hidden in the black box are revealed.

Note that the fact that mechanisms are described as causal pathways does not entail that they are linear causal chains. For example, in the absence of measurements of intermediary stages, a cyclical metabolic pathway, such as Krebs' cycle, appears as a linear input–output phenomenon (2 acetyl-CoA, 6 NAD^+, 2 FAD, 2 ADP+P_i $\rightarrow$ 4 CO_2, 6 NADH, 6 H^+, 2 $FADH_2$, 2 ATP, 2 CoA). The same applies to the mechanism of T-cell activation described in Section 1. Even though the phenomenon to be explained is a stimulus–response behavior (Figure 1), the molecular mechanism of this phenomenon includes a negative feedback loop, ensuring that inflammatory responses eventually shut down (Figure 2).

In summary, a phenomenon is analogous to an incomplete or low-resolution measurement-mediated representation of a mechanism. The causal interpretation of measurements dictates that the mechanism causally determines the phenomenon for which it is responsible. This causal relationship, however, should not to be understood in the etiological sense of bringing about changes in a variable, but rather in that of linking measured variables in relationships of association and causal dependency or, in more general terms, not allowing them to vary independently of one another.

3.5 A Level-Free Conception of Mechanisms

In their analysis of the relationships between mechanisms and phenomena, Craver and Bechtel (2007, 549) rely on two distinct notions of constitution: a noncausal one between physical systems (e.g., the eye being composed of cells); and one between behaviors, or phenomena, which is amenable to a causal interpretation (e.g., rhodopsin signaling pathway is part of light-to-neural-activity transduction). They further assume that behavioral and physical constitutions are systematically aligned, such that the behavior of a system is hierarchically composed of and experimentally decomposable into the behaviors of its parts (e.g., the light-transducing eye consists of rhodopsin signaling cells). Finally, they argue that the physical constitution element of this hierarchy rules out a causal relationship between mechanisms and phenomena (hence, the light-transducing eye is not *caused by*, but instead is *made of* rhodopsin signaling cells).

In contrast, the causal mediation account neither assumes nor entails that reality is systematically structured as hierarchies of ontologically distinct macro-state phenomena consisting of micro-level mechanistic components. On this account, one and the same causal structure is probed by an increasing number of measurements and interventions gradually revealing additional causal intermediaries. The outputs of measurements are taken to be physical effects, and in this sense are ontologically distinct from the measured mechanisms; however, they are not made of underlying or lower-level mechanisms.

This divergence requires closer scrutiny. Why assume that behavioral and physical constitutions are systematically aligned? And why is the latter an essential ingredient of constitutive relevance? I have reasons to believe that neither assumption is justified. A lack of alignment between behavioral and physical constitutions is not that uncommon. For example, erythema is consistently induced in the skin of many organisms by a variety of harmful stimuli, including radiation, microbial infection, insect bites, tissue damage, as well as chemical compounds such as prostaglandins, liposaccharides, and cytokines. Depending on the intensity/concentration and the targeted part of the body, these very same stimuli are also known to consistently induce fever, pain, and loss of function. Since all these phenomena partially share the same mechanistic structures, they may be said to be constitutive of a general phenomenon of inflammation. For instance, stimulus-induced synthesis of prostaglandins is constitutive of both stimulus-induced local erythema and stimulus-induced systemic fever. Behavioral constitution, however, is not paralleled by experimental setups standing in part–whole relationships. All of the above phenomena are documented in very similar experimental setups involving whole organisms.

Still, it may be objected that prostaglandins bind cell membrane receptors in cell-free systems and that binding triggers a signaling cascade, altering gene expression in cell models. This suggests a hierarchical structure of experimental setups roughly matching the organismic, cellular, and molecular levels of composition. True, but this should not distract us from the fact that the crucial piece of evidence here is that receptor binding and gene expression causally mediate inflammatory responses. Researchers didn't infer that gene expression is part of the mechanisms responsible for inflammatory responses because gene expression occurs in cells, and cells are parts of organisms, but because receptor binding and gene expression are causal intermediaries along the causal pathway linking exposure of an organism to a harmful stimulus and the ensuing inflammatory response. To be sure, part–whole constitution provides important information about the physical world – I will come back to this in Section 4 – but it is simply not true that top-down and bottom-up interventions are experiments in which the independent and dependent variables stand in part–whole constitution relationships. From an experimental standpoint, there is no such thing as intervening on or measuring cells or organisms, but only intervening on or measuring specific variables in the context of experimental setups involving cells or organisms.

A knockout-type experiment can demonstrate causal mediation irrespective of considerations about physical constitution. The experimental design behind such experiments relies on a strictly contrastive notion of causation, which is insensitive to whether the causal connection is 'fine' or 'coarse-grained' (i.e., how far apart along a causal pathway are the points of intervention and measurement), or whether the variables involved are deemed 'lower' or 'higher-level' (Baetu 2017b; Kendler and Campbell 2009; Woodward 2008). If it is possible to intervene on one variable and measure another, then an experiment testing for causal relevance or mediation can be conducted. There is nothing methodologically problematic about aggregating seemingly higher-level determinants of erythema, such as exposure to sunlight or tissue damage, with lower-level ones, such as prostaglandin synthesis, as elements of the same causal structure. In fact, it is even possible to aggregate biological, psychological, and social determinants of an outcome, say pain intensity, in what is commonly referred to as a 'biopsychosocial model'. The implication here is that mechanisms, as elucidated in the experimental practice of biological sciences, are by default neutral to debates concerning levels and inter-level causation.

3.6 The Completeness of Mechanistic Explanations

The causal mediation account also implies continuity between descriptions of phenomena and mechanistic explanations. The two don't refer to ontologically distinct categories. Rather, both the description of a phenomenon and that of its mechanism refer to the same causal process, except that a phenomenon is described in light of measurements probing very few variables, thus appearing as a black box or coarse-grained association of variables, while a mechanistic description is based on measurements ideally describing all relevant variables and their dynamics, revealing what was initially hidden in the black box. This conclusion is consistent with the view that the elucidation of mechanisms is a matter of gradually filling in missing mechanistic details (Craver and Darden 2013, Ch. 10; Darden 2006b, Ch. 4; Machamer et al. 2000).

If this is the case, then what distinguishes a complete mechanistic description from an incomplete one?[15] Intervention experiments, including knockout-type

---

[15] Craver and Bechtel (2007, 549) answer this question as follows: "Each new decomposition of a mechanism into its component parts reveals another lower-level mechanism until the mechanism bottoms out in items for which mechanistic decomposition is no longer possible." For this answer to work, behavioral and physical constitution must be systematically aligned, such that when physical constitution bottoms out there are no behaviors left that may constitute the phenomenon of interest. But this is not the case. As exemplified earlier, behaviors can constitute other behaviors irrespective of levels of physical composition.

ones, can only demonstrate causal relevance given a causal background common to test and control conditions. Different experiments test different factors, bringing into focus what other experiments leave as part of the background. It is reasonable to expect that causal models constructed by piecing together experimentally demonstrated causal relationships succeed in providing substantial knowledge about the mechanisms responsible for a phenomenon. Nevertheless, this experimental design does not provide criteria for assessing whether all the relevant factors have been identified or whether the description of the mechanism is complete.

This led some authors to argue that the completeness of a mechanistic explanation is a question of pragmatic perspective. There are many ways in which a mechanistic explanation is useful to the scientist, and the purpose will determine when the explanation is deemed to be complete (Craver and Darden 2013, Ch. 6). For instance, if one seeks to gain control over the phenomenon of interest, then the explanation should focus on manipulable factors, such as gene and protein sequences. If the main goal is prediction, then the emphasis will fall on showing how changes in variables describing a mechanism result in changes in some outcome of interest. If one seeks intelligibility, be it for didactic purposes or in order to reveal general patterns, then abstraction or even idealization may be necessary.

In parallel with the pragmatic stance, there is also a more objective sense in which mechanisms can be shown to be complete. The fact that, at least in some cases, mechanistic components can be physically extracted from and inserted back into a system suggests that a mechanism can be reliably decomposed into a set of parts and then recomposed back starting from these parts. This is routinely achieved for genetic determinants by combos of knockout-transgenic experiments. Another example is the production of erythema exudates in an ultraviolet-induced erythema model of inflammation, which are then collected and injected in the skin or paw of a rodent in order to produce a pain response. A more ambitious experiment is the in vitro reconstitution of a whole molecular mechanism (Craver and Darden 2013, 92–94; Weber 2005, 106–08). Such an experiment demonstrates that mechanistic components having the properties and organization specified in the explanation are sufficient to produce the phenomenon of interest in the absence of the physiological context of a cell, organism, or other biological system, thus demonstrating that no other physical components are needed. The mechanism itself, as a causal process, is deemed complete in the sense that an experimental setup has been successfully purified down to a minimal physical system still capable of producing the phenomenon. The mechanistic description is complete in the sense that it includes evidence that no other mechanistic components are required. By combining these two notions of completeness, we obtain a complete mechanistic explanation stating that mechanistic components having the properties and organization specified in the mechanistic description are all that is needed to produce the phenomenon (Baetu 2015b). Alternatively, synthetic biology (Morange 2009), mathematical models of mechanisms (Baetu 2015a; Bechtel and Abrahamsen 2010; Braillard 2014; Gross 2015; Issad and Malaterre 2015), as well as the ability to correctly foresee and correct side-effects of treatments and other technological applications based on mechanistic explanations may also provide evidence for completeness.

4 What Is the Physical Nature of Biological Mechanisms?

The account of phenomena and the causal mediation account of the relationship between mechanisms and phenomena defended in the previous sections assume a minimalist ontology, which best describes experimental research conducted outside any discipline-specific theoretical framework. Methodological assumptions about the nature of reality are limited to the expectation that differences in measurements are informative of differences in the causal structure of reality. There is no point in agonizing over what exactly this structure is and how measurements reflect it. Experimental methodology doesn't have the resources for addressing these questions. It only asks us to accept that if the same technique yields two different measurements on two occasions, then there is something different about reality on these two occasions. Gaining additional knowledge requires theorizing, formulating hypotheses about how reality and experimental techniques work, which provide the ingredients for the more substantial ontologies typically endorsed by biologists. The goal of this section is to provide an overview of the main ontological accounts developed in the philosophical literature, along with a critical assessment of the experimental, theoretical, and metaphysical rationales justifying these accounts.

4.1 Beyond the Minimal Experimental Interpretation

The minimal experimental ontology consists of variables and relationships between variables. It would be misleading to think of variables as physical entities or structures and of relationships between variables as activities, interactions, or functions. Inasmuch as nothing else is known or assumed by hypothesis, variables are defined in operational terms, according to the techniques used to measure and manipulate them. This leaves unanswered questions about the physical nature of variables, as well as about potential identities, part–whole composition, and other physical relationships between variables.

Agnosticism about physical nature extends to the notions of association and causal and constitutive relevance, which are generic relationships between variables satisfying relatively broad desiderata. A point raised in the previous section is that there are no levels restricting relationships between variables. James Woodward (2008, 222) argues that, from the perspective of an interventionist account of causation, nothing justifies giving "automatic or a priori preference to any particular grain or 'level' of causal description over any other" and "there is no bar in principle to mixing variables that are at what might seem to be different 'levels' in causal claims." If controlled intervention experiments are all that is needed to justify claims about causation in empirical science, and if the same method of controlled interventions is successfully applied in all cases, then there are no reasons why some causal determinants should be assigned a distinct status. This methodological argument can be further generalized. For as long as there are no reasons to suspect flaws in experimental design, any correlate or causal determinant of a given outcome, as well as any causal intermediaries along input-specific pathways leading to that outcome can be aggregated in the same network of correlations, causal determinants, and causal pathways linking variables (Baetu 2017b). There are no a priori set limits to how far such networks can extend and which variables they may include.

The only departures from this otherwise austere – or, from a different perspective, very liberal – metaphysical picture are marked by the interpretation of some variables as physical factors and organizational features suggested by spatiotemporal details such as the localization of variables within the spatiotemporal boundaries of experimental setups, dynamic features, and the ordering of variables along causal pathways. These specifications anchor variables to the physical reality of an experimental setup.

Some may be tempted to dismiss this minimalistic ontology as a mere methodological idealization. After all, no aspect of scientific practice is ever conducted in a state of theoretical vacuum. The objection is certainly valid. As a matter of fact, biologists tend to be realists committed to richer ontologies. Nevertheless, it would be a mistake to underestimate the importance these very same biologists attach to the fact that shared methodological principles drive research across the life sciences, basic and clinical. Whatever is hypothesized on theoretical grounds must eventually be shown to be causally relevant to a biological outcome, and the gold standard for producing such evidence dictates that controlled experiments be conducted. The requirement of empirical confirmation alone constitutes a serious obstacle to the notion that one could simply ignore the constraints experimental methodology imposes on mechanistic ontology.

Furthermore, evidence for causal relevance functions like a common currency, making it possible to aggregate knowledge from different sources. To give a few examples, the diagnosis and pathology of psychiatric conditions encompasses a mixture of physiological and psychological symptoms, risk factors, and causal determinants (Kendall 2001; Kendler and Campbell 2009; Murphy 2006; Schaffner 1993); epidemiological studies reveal that many common diseases are multifactorial and can be influenced by factors such as genes, diet, life style, and socioeconomic status (Broadbent 2013, Ch. 10); while pain "is determined by the interaction among biological, psychological (which include cognition, affect, behavior), and social factors (which include the social and cultural contexts that influence a person's perception of and response to physical signs and symptoms)" (Asmundson and Wright 2004, 42). Evidence for causal relevance also marks an important point of intersection between basic science and clinical research. The latter is deliberately pragmatic, amounting to the application of methodological principles for clinical purposes (Hill 1965). Evidence-based medicine further promotes a scientific practice systematically sanitized of theoretical elements for fear that reliance on theoretical rationales in the absence of direct experimental evidence may lead researchers astray, with disastrous consequences for medical practice (Howick 2011). Thus, interdisciplinary research, especially when it involves interactions between basic and clinical science, presupposes a minimal experimental interpretation as a common basis for sharing, aggregating, and applying findings.

It is also possible that some readers may lean heavily in the opposite direction, favoring a radical empiricism according to which a minimal experimental interpretation is all that science can justifiably hope to achieve, dismissing anything over and above this interpretation as futile metaphysical speculation. Adopting such a stance would also be a mistake. There is more to science than experimental methodology. Richer representations of mechanisms are made possible by taking into account theoretical knowledge about the physics and chemistry of biological systems. In many cases, this knowledge is put to work under the form of hypotheses and assumptions about what is happening during measurement and intervention procedures. Most of the techniques routinely employed in cell and molecular biology – including microscopy, electrophoresis, radiolabeling, chromatography, and crystallography – were developed based on theoretical considerations from physics and chemistry. The general principles governing these techniques are well understood, and researchers have a fairly good idea about what is going on during experimental procedures. Thanks to this theoretical knowledge, we have reasons to believe that biological mechanisms do not look and behave like eighteenth-century automata, but are much more complex, stochastic, and noisier systems composed of myriad nonrigid parts; that the causal dependencies between parts are often back-and-forth equilibria; and that their organizational features are seldom fixed, but both depend on and determine the interactions between the various parts of the mechanism.

Something needs to be said about these richer representations of mechanisms and the ontologies they entail. A thorough discussion of the theoretical foundations of biology and its experimental techniques is beyond the scope of this short Element. Marcel Weber (2005), Lindley Darden (1991; 2006b), William Bechtel (2006; 2008), Carl Craver (2007), Frederic Holmes (2001; 2006), and many others analyze specific episodes of scientific discovery, examining in more detail the theoretical background and specific experimental techniques that shaped our current conception of mechanisms. What follows is a general assessment of the mechanistic ontologies developed in the philosophical literature, with a focus on how they relate to scientific theories and experimental evidence, as well as the extent to which they diverge from the minimal experimental interpretation.

4.2 Part–Whole Composition

The constitutive interpretation proposed by Carl Craver and William Bechtel (Section 3.3) adds a novel element to the experimental interpretation: at least some variables are associated with physical objects standing in part–whole composition relationships. In turn, the association of variables with parts of an experimental setup provides the empirical basis for characterizing mechanisms as systems of parts, which is undoubtedly a key element of the concept of mechanism.

As discussed previously, I don't believe the localization of variables to spatiotemporal parts of an experimental setup licences the conclusion that parts are variables targeted by measurements and interventions, as this results in methodological inconsistencies. I also don't think there are sufficient grounds to justify the claim that part–whole physical composition is systematically matched by behavioral (phenomenal) composition, as postulated by a hierarchical model of levels of mechanisms. These points of divergence aside, I agree with Craver and Bechtel that physical decomposition and recomposition, such as isolating and transferring parts from one experimental setup to another, are widely used methods for localizing causes responsible for differences in outcomes. An example mentioned previously is the localization of the biochemical compounds mediating inflammatory responses. By injecting exudates from lab animals exposed to ultraviolet radiation in the skin of animals with no history of inflammation it was established that something in these exudates causes an inflammatory response. Since compounds such as prostaglandins can be extracted from and put back into erythema exudates, the former are said to be physical parts – as opposed to just being located within the spatiotemporal boundaries – of the latter. The same applies to erythema exudates in relation to animal organisms and to organisms in relation to the experimental setups used to study the phenomenon of ultraviolet-induced erythema.

There are parts and wholes in this picture, although this is not enough to justify claims about levels of composition. Levels imply some degree of generality, while the above-mentioned experimental practices only describe how particular experimental setups are decomposed for the purposes of a handful of experiments. A more robust notion of levels emerges if we further consider the fact that the manipulation and measurement of multiple variables involve the same decomposition techniques. For example, since protein, DNA, and RNA purification require that cells be lysed and that a soluble fraction be separated from a membrane pellet, the three may be said to belong to the same level of composition. Ina much as decomposition techniques are an integral part of other decomposition techniques – e.g., tissue homogenization followed by cellular lysis followed by fractional centrifugation in order to separate cytoplasmic contents from the cytoplasmic membranes – it may be argued that some variables belong to 'higher' or 'lower' levels of composition than others. In turn, a layering of variables according to shared decomposition techniques may justify claims about interventions on variables at one level having an effect on variables at another level (Bickle 2006).

It seems unlikely, however, that experimentally defined levels of composition coalesce into the familiar molecules-organelles-cells-tissues-organs-organisms hierarchy or some other general pattern of organization (Potochnik and McGill 2012). Take, for instance, erythema exudates. These accumulations of interstitial fluid belong to a local hierarchy, being parts of organisms composed of molecules, but are not cells, tissues, or organs. Other variables don't seem to belong to any hierarchy at all. Ultraviolet radiation, for example, is part of an experimental setup, but is neither higher, nor lower level than guinea pigs, erythema, or prostaglandins. Even accepted notions, such as that of a molecular level, are in fact quite vague. A distinction can be made between the level of composition proper to molecular biology explanations involving molecular mechanisms, such as signal transduction pathways; the level of biochemical explanations of metabolic activities, such as glycolysis; and the level of physical chemistry explanations, such as structural explanations of the affinity of an enzyme for its substrate (Morange 2002). These inconsistencies suggest that a universal notion of levels should be abandoned in favor of a local one, as some authors have recently proposed (Brooks 2017; Craver 2007; Love 2012).

4.3 Properties, Interactions, and Causal Transmission

The localization of variables to physical parts of experimental setups further suggests a 'properties and interactions' ontology. Localized variables correspond to aspects of the physical reality of parts that we may refer to as 'properties of parts'. Causal dependencies between variables associated with different parts may be viewed as causal interactions between parts. This terminological glossing marks potential points of intersection between experimental results in biology and a theoretical understanding grounded in chemistry and physics. For instance, the parts may turn out to be molecules described by variables referring to properties such as atomic composition and molecular geometry, while causal dependencies between these variables may reflect chemical interactions involving bonding and atom exchanges.

A physical interpretation marks a crucially important shift from the largely agnostic stance associated with the minimal experimental interpretation to the more substantial realism typically espoused by both scientists and the new mechanistic philosophers (Craver and Darden 2013; Illari and Williamson 2011). A difference-making notion of causation glues together variables into more-or-less detailed descriptions of phenomena and mechanisms. Aside from satisfaction of the methodological desiderata of experimental control, there are no rules, constraints, or any other kind of limitations on which variables may stand in causal relationships. Interactions, on the other hand, glue together physical objects and are heavily constrained or even strictly determined by the laws and principles at the heart of accepted theories and explanations from physics and chemistry.

William Wimsatt (2007, 204) assumes a 'properties and interactions' interpretation when he characterizes levels of composition as "families of entities usually of comparable size and dynamical properties, which characteristically interact primarily with one another, and which, taken together, give an apparent rough closure over a range of phenomena and regularities." According to this view, mechanisms are no longer mere webs of causal dependencies, as argued by Woodward (2002), and not even causal pathways involving variables localized within specific experimental setups, as I myself suggested in Section 2. Rather, they are physical parts engaged in physical interactions. This interpretation is reflected in some of the most influential characterizations of mechanisms. Most notably, Stuart Glennan defines a mechanism as "a complex system which produces that behavior by the interaction of a number of parts according to direct causal laws" (1996, 52) or "direct, invariant, change-relating generalizations" (2002, S344).

There is some flexibility regarding the exact nature of physical interactions and the principles governing them. Unlike Cartesian mechanisms, tightly linked to the theory of classical mechanics and the clockwork view of the world, biological mechanisms may involve different kinds of interactions, from push–pull mechanical forces to chemical bonding and thermodynamic processes (Machamer et al. 2000). It is even possible that some biological phenomena exploit quantum mechanical effects (Ball 2011; Barwich 2015).

At the same time, the requirement for physical interaction also imposes constraints on the kinds of parts that may interact with one another. For instance, it has been argued that psychological and social factors cannot be integrated in biological mechanisms based solely on evidence demonstrating their causal relevance to a common outcome. They must also be shown to have a biological basis in virtue of which they interact with the biochemical machinery of a cell or organism. From an interactionist standpoint, models of psychiatric conditions mixing neurological and psychological considerations, as well as multifactorial models of disease aggregating biological and socioeconomic risk factors are, at best, incomplete explanations (Frith 1992, Ch. 3; Gori 1989; Krieger 1994). According to this view, webs of causal difference-making dependencies require a more detailed explanation. Mechanisms provide such an explanation by showing how coarser-grained causal regularities are mediated by more fundamental chemical and physical interactions between molecular parts.

This interpretation suggests a metaphysical picture in which biological mechanisms occupy an intermediary place between the more fundamental principles of physics and chemistry, and the exception-ridden regularities typical of biology, such as the genetic code, Kleiber's law, or developmental generalizations like 'all snakes lack forelimbs' (Baetu 2012b; Glennan 1996). The notion that a complete mechanistic explanation must ultimately describe physical interactions between parts also forces a shift in the way causation is conceptualized: coarse-grained, difference-making causal dependencies are ultimately underpinned by finer-grained, physical interactions involving an exchange of marks (Salmon 1984, Chs. 5–6) or conserved quantities (Dowe 1995; Salmon 1997). A mark-transmission causal concept is supported by the widespread use of experiments involving labelling (e.g., radiolabelling, immunolabelling, fluorescent tagging) and binding assays (e.g., two-hybrid screening, electrophoretic mobility shift, chromatin immunoprecipitation). These experiments are explicitly designed to demonstrate physical interaction, usually for the purpose of elucidating intermediary mechanistic stages, by tracking atom exchanges between molecules or identifying new mechanistic components using known molecules as 'binding bait'.[16] A physical quantity-transference notion of causation is supported by a theoretical line of reasoning claiming that biological explanations are likely to bottom out at a molecular level because at this size scale the mechanical, electrostatic, chemical, and thermal energies have similar magnitudes (Philips and Quake 2006). The convergence of energy values means that these various forms of energy can be interconverted, which may explain some of the most fundamental properties of living things, such as their ability to convert food (chemical energy) into motion (mechanical energy). Furthermore, thermal energy (random molecular collisions) may literally 'kick start' molecular mechanisms, thus explaining their ability to work autonomously and spontaneously. In contrast, mechanical forces at macroscopic scales or atomic binding forces stabilizing molecules are largely insensitive to thermal fluctuations, rendering impossible phenomena such as spontaneous activation, self-assembly, or change of shape (Hoffmann 2012, Ch. 4). Thus, neither atoms nor macroscopic objects, but only molecules have the right kind of physical properties that can explain the fundamental characteristics distinguishing living things from other physical systems. This rationale vindicates a long-standing reductionist intuition that there is something objectively special about molecular explanations in biology (Bickle 2006; Rosenberg 2006).

Finally, it is opportune to mention Glennan's (2010, 260) recent discussion of 'ephemeral mechanisms' amounting to "collection[s] of interacting parts where: 1. the interactions between parts can be characterized by direct, invariant, change-relating generalizations; 2. the configuration of parts may be the product of chance or exogenous factors; 3. the configuration of parts is short-lived and non-stable, and is not an instance of a multiply-realized type." The intuition here is that regular mechanisms as well as rare mechanism-like processes may be predicted based on an understanding of how the parts of a system interact with one another. The early twentieth-century 'bag of enzymes' model of the cell illustrates this idea. According to this model, the cytoplasm of a cell is an unstructured aqueous solution containing all the enzymes and substrates required for cell metabolism. The model postulates

---

[16] For more detailed examples, see 'forward' and 'backward chaining' in (Craver and Darden 2013, Ch. 5; Darden 2006b, Ch. 3). Note that a difference-making concept of causation is still present. Labelling and binding assays rely on control conditions necessary to demonstrate specificity of binding under physiological-like conditions. For instance, labelling must be shown to specifically target only the molecules of interest, and not indiscriminately tag a large number of molecular species. Equally important, evidence of binding, atom exchange, or some other form of interaction does not guarantee biological relevance. In order to play a role in biological explanation, the interaction must further be shown to be a difference-maker with respect to a biological outcome. For a discussion of the issue of biological relevance in the Meselson-Stahl labeling experiment, see (Baetu 2017c).

that the spatiotemporal organization of molecular mechanisms, from the directionality of metabolic pathways to the assembly of supramolecular structures, is driven by specificity of binding in the context of a free diffusion chemistry. In other words, were it not for the differential binding specificities among various molecular species, a cell would display the same amount of order as sugar dissolved in a glass of water. One strength of this admittedly idealized model is that it applies to both cells and the prebiotic 'organic soup' from which life might have emerged. The difference concerns primarily the concentration and variety of molecules. While the cytoplasmic membrane keeps all the molecules needed for cell metabolism at an optimal concentration, prebiotic environments were presumably much more diluted and less complex, and for this reason displayed what one may describe as a chaotic and sluggish metabolic-like activity. Considered from this perspective, there is indeed something ephemeral about molecular mechanisms. They are not rigid, permanent structures analogous to everyday mechanical devices. Instead, spatiotemporal organization, macromolecular structure, and the frequency with which molecular mechanisms operate are determined by chemical interactions between parts, changing from permanent to fleeting and from regular to exceedingly rare depending on the number of parts and the presence or absence of certain organizational constraints.

4.4 Physical Identities

Browsing protein databases reveals an interesting fact: a significant number of proteins are known by two or three different names. Duplications occurred because many proteins were initially identified and characterized in the context of research projects aiming to explain different phenomena, and it is only with the advent of large-scale sequencing and enhanced techniques of genetic engineering that it became evident that they share the same or significantly overlapping sequences, are products of the same genes, and their mutation or knockout affects several traits.

Identities of this sort indicate that the mechanisms responsible for distinct phenomena share components and therefore may not be entirely modular. For instance, it turns out that intracellular signaling pathways and gene regulatory mechanisms intersect at multiple points, forming widespread molecular networks (Davidson and Levine 2005). Biologists always expected that molecular mechanisms are subject to interferences due to nonspecific binding between molecules, which may be thought of as a more-or-less constant background noise affecting all molecular mechanisms. Whether this source of interference is something cells can exploit or merely a nuisance they need to minimize is still a matter of debate. Shared molecular components further generate specific patterns of interference selectively affecting some mechanisms but not others, and therefore are more likely to be biologically significant.

It is less clear how physicalist models postulating systematic identity (Smart 1959), supervenience (Kim 2005, Chs. 1–2), or constitutive (Craver 2007) relationships between psychological and biological states fit within a mechanistic framework. Despite their popularity in philosophy of mind, none are directly supported by experimental results. As far as neuroscience is concerned, there is ample experimental evidence demonstrating that biological variables are causally relevant to psychological outcomes and that at least some psychological variables have an effect on biological outcomes (e.g., placebo effect, psychosocial determinants of pain). It is therefore not particularly surprising that proposed biological explanations of psychological phenomena have a lot to do with causal mechanisms (Koch et al. 2016) and very little with noncausal relationships (Hohwy 2007).

One potential empirical argument for psychoneural identities may exploit well-documented cases of psychological measurement techniques (self-report, cognitive tasks, behavioral descriptions) being replaced by biological ones (neurological and physiological correlates, molecular markers). Nevertheless, replacement of one technique by another does not demonstrate the identity of the variables targeted by these techniques. Replacements are invariably justified by validation procedures whereby biological tests are shown to consistently give the same results as psychological tests. Thus, biological tests are in fact extensions of earlier psychological tests motivated by the discovery of biological correlates and determinants of psychological outcomes. The use of cortisol as a biomarker of psychological stress (Smyth et al. 1997) illustrates this point.

4.5 Structures and Capacities

According to a dispositional interpretation, phenomena are capacities describing how a system responds in certain conditions, such as the presence of a stimulus (Cartwright 1989; Cummins 1975; 2000). Just like behaviors, the capacities of wholes are analyzed in terms of capacities of their parts. For instance, the capacity of prostaglandins to bind their receptors may be said to contribute to the organism's capacity to develop inflammation in response to harmful stimuli.

According to some analyses, capacities are ultimately grounded in the physical structure of a system, namely the spatiotemporal arrangement of its parts and the bonds keeping these parts in place, which must be present in order for a stimulus to induce a response (Ylikoski 2013). The classical example in philosophical discussions is fragility or, in the terminology of materials science, brittleness. Suppose two materials, a metal and a glass sheet, are exposed to a stress, resulting in a superficial crack of comparable magnitude. Why does glass break while metal doesn't? The textbook explanation tells us that metal is made of layers of atoms sliding along one another, thus blunting the deep end of the crack and preventing it from advancing. In glass, however, atoms are partially held together by ionic bonding. Sliding would result in an alignment of ions of the same charge, which costs energy. As a result, atoms don't slide and, instead of being blunted, the crack keeps advancing until glass eventually fractures.[17]

The emphasis on physical structure counterbalances the requirement for physical interaction. For example, the notion of 'specific binding' is reminiscent of a capacity, which, just as brittleness, is made possible by a certain chemical makeup and internal bonding. For example, the claim that the capacity of prostaglandins to bind their receptors contributes to the organism's capacity to develop an inflammatory response is backed up by experimental interventions, such as blocking receptor binding using antibodies, which results in a lesser or absent inflammatory response when the organism is subjected to a harmful stimulus. The evidence behind the claim that the capacity of two molecules to bind one another is made possible by atomic composition and molecular geometry is more complex. The discovery of isomerism revealed that chemical and physical properties are not determined by atomic composition alone. Early evidence obtained by comparing the properties of compounds characterized by identical proportions of elementary atoms suggested that the missing ingredient is molecular geometry. This suggestion was eventually confirmed by more direct evidence showing that amino-acid substitutions and other molecular modifications result in conformational (geometrical) changes in molecules, as demonstrated by X-ray crystallography and other imaging techniques.

---

[17] I am not sure what to make of the modal glossing associated with the notion of capacity. The scientific explanations used to illustrate claims about dispositional capacities are invariably descriptions of what happens when a system is actually exposed to a stimulus (Kaiser and Krickel 2017, 765). As a case in point, the above explanation of brittleness is a description of what happens when a material is actually subjected to a stress. Moreover, the phenomenon explained here is not an inherent propensity of a material, but rather the difference in behavior when different materials are subjected to the same stress. Likewise, chemical affinity is not an intrinsic dispositional property of any given molecule, but a measured rate with which a molecular species actually binds another species in a given chemical environment. Since any two molecular species have some binding affinity, what is explanatorily relevant are relative differences in affinity. For example, prostaglandins have a higher affinity for their receptors under physiological conditions (i.e., prostaglandins bind these receptors more often or any given prostaglandin molecule has a higher probability of binding a receptor molecule) than for other molecular species also present under physiological conditions.

In addition to chemical structure, there are also various forms of biological structure. The free diffusion assumption underlying the 'bag of enzymes' model mentioned earlier has recently been challenged by pointing out that molecular mechanisms operate within an intracellular environment filled with other molecules. Macromolecular crowding functions as an excluded volume effect, favoring the aggregation of macromolecules while drastically decreasing the rate of any diffusion-dependent molecular activity (Ellis 2001). This strongly suggests that the intracellular environment cannot consist of a disordered collection of molecules, as this would result in the proliferation of nonspecific interactions at the expense of order-generating specific ones. Instead, it must be structured in such a way as to bring in close proximity proteins and their ligands, thus favoring the specific chemical interactions required for the operation of molecular mechanisms (Hochachka 1999; Mathews 1993). If this is the case, however, then the tridimensional structure of the cell and tissue organization are an important source of order, functioning as a scaffold constraining the behavior of molecular mechanisms by favoring some chemical interactions while suppressing others.[18] This explanation too relies on a notion of physical structure, albeit not that of individual molecules, but rather of the intracellular environment.

Finally, it has been argued that stable physical structures are borderline mechanisms (Illari and Williamson 2012, 130–31). For instance, a double covalent bond between two carbon atoms may be viewed as the mechanism responsible for keeping the functional groups attached to each carbon atom in a fixed position relative to one another, not allowing them to rotate or come in closer proximity. In turn, a rigid arrangement of functional groups may be part of the mechanism responsible for the specific binding between two molecules. Even if one disagrees that such structures are by themselves mechanisms, structural features are often shown to be constitutively relevant to biological mechanisms in the sense that the absence or modification of these features has an impact on biological phenomena.

The explanatory relevance of physical structure suggests a very interesting metaphysical picture that has the potential of diverging significantly from the minimal experimental interpretation. It may be argued that phenomena cannot be indefinitely analyzed in terms of causal mediation. There is a point beyond which interventions are no longer possible and a causal interpretation is no longer on firm experimental grounds. The maximum level of detail with which biological mechanisms can currently be probed is via experimental

---

[18] Recent evidence supports this conclusion. See, for instance, the effects of nuclear architecture on transcription (Cremer and Cremer 2001).

interventions targeting parts of molecules, such as monomeric components of polymers and functional groups of organic molecules, which can be modified via genetic engineering and chemical synthesis. Such interventions are known to result in conformational changes, altered binding affinity and chemical reactivity, as well as differences in biological outcomes. Up to this point, knowledge about the composition of matter from chemistry and physics is paralleled by intervention techniques demonstrating the causal and constitutive relevance to biological outcomes and phenomena.

Beyond this point, however, as we approach the quantum divide marking the boundary between the 'micro' and 'macro worlds', there is a mismatch between theory and intervention. For instance, the valence shell electron pair repulsion theory allows chemists to predict the shape of molecules based on measurements of electron density, but the theory is not paralleled by intervention techniques targeting electrons. The very notion that electrons could be localized and individuated is at odds with quantum mechanics (Hendry 2008). It is certainly claimed that electron density determines molecular geometry, but in this context 'determines' refers to what theories about the behavior of electrons entail with respect to the properties of molecules, and not to an experimentally demonstrated relationship of causal relevance. In the absence of the latter, it would seem that at least some fine-grain details of mechanistic descriptions are not amenable to a causal interpretation, although it is difficult to say whether this justifies mere agnosticism or a commitment to a noncausal interpretation.

4.6 Functions, Abstract Models, and Biological Significance

Molecular biology textbooks routinely compare molecular mechanisms with manmade machines performing specific functions. Take, for instance, the following passage from a popular introductory book: "molecular machines have many similarities with familiar machines like scissors and automobiles. The unusual, organic shapes of molecular machines may seem daunting and incomprehensible, but in many ways, molecular machines may be understood in a similar way: as a *mechanism where parts fit together, move, and interact to perform a given job*" (Goodsell 2009, 9, emphasis added). Given the abundance of functional ascriptions in the biological literature, William Bechtel and his collaborators popularized the view that mechanistic explanations "propose to account for the behavior of a system in terms of the functions performed by its parts and the interactions between these parts" (Bechtel and Richardson 2010, 17). Or, according to an alternative formulation, "a mechanism is a structure performing a function in virtue of its component parts, component operations, and their organization. The orchestrated functioning of the mechanism is responsible for one or more phenomena" (Bechtel and Abrahamsen 2005, 423).

The fact that functional characterizations of mechanisms draw on analogies with manmade machines raises the possibility that functional ascriptions are not necessary elements of scientific explanation, but rather analogies meant to enhance intelligibility. Nevertheless, this cannot be the whole story. Functions can be understood more rigorously as generic relationships, especially mathematical ones, which may approximately hold true irrespective of the particular nature of the variables serving as relata. A mechanism may therefore be viewed as an abstract model describing functional relationships between variables associated with the different parts of a system. It could be an analogical model, in which case its value is heuristic or pedagogical, or it could be a mathematical model capable of generating rigorous, quantitative predictions. For instance, given some additional tweaking, the dynamic behavior of the regulatory mechanism depicted in Figure 2 can be accurately modeled using the general mathematical template employed to describe an oscillator (Hoffmann et al. 2002). This said, a conception of mechanisms as abstract models is meant to remain epistemic, unless the general relationships around which the model is built are construed as law-like regularities embodying some form of physical necessity.

Functional ascriptions are also common in specialized molecular biology and genetics journals, although not as a form of abstract modeling, but rather in connection with claims about 'biological significance' and 'biological information'. In biochemistry, mechanisms, such as metabolic pathways, are described as series of chemical reactions characterized in terms of reactants, products, transition states, energy requirements, and reaction kinetics. This is consistent with the properties and interactions ontology discussed in Section 4.3. Molecular biology, on the other hand, describes mechanisms in terms of the biological significance of molecular components and interactions, while ignoring many fine-grained chemical details such as transition states and quantitative-dynamic aspects of chemical interactions; see, for instance, the description of the NF-κB regulatory mechanism in Section 1.1.[19] The guiding intuition is that biological significance, or causal difference-making to a biological outcome as established by means of knockout and reconstitution-type experiments, makes possible a functional decomposition of the behavior of the system (the phenomenon to be explained) into functions or operations of parts and their interactions. Thus, when experimental interventions result in the

---

[19] Recent research favors an integration of molecular and biochemical descriptions, for instance, by using differential equations to model quantitative and dynamic aspects of chemical reactions taking place in molecular mechanisms (Baetu 2015a; Bechtel and Abrahamsen 2010).

loss or reduction of an outcome, such as the catalytic activity of an enzyme, the phenomenon of interest is altered in a way often described in the scientific literature as a 'loss of function'. Conversely, when interventions restore the outcome, or result in an increased value of the outcome or novel outcomes, absent in the control condition, there is a 'regain' or 'gain of function'.

When experimentally induced disruptions of molecular mechanisms result in a failure of cells and organisms to respond to environmental cues, such mechanisms are often referred to as 'signaling pathways'. For instance, molecular explanations of inflammatory responses don't focus on the physics and chemistry involved in the binding between prostaglandins and their receptors, stressing instead the biological function or significance of prostaglandins as 'signaling' or 'carrying information about' the presence of a threat to the organism. A similar comment applies to the notion of 'genetic information'. The mechanisms of transcription and translation may be described as a series of chemical reactions, with DNA on the side of reactants and polypeptides on the side of the products. In molecular biology, however, the preferred description is in terms of the difference the nucleotide composition of a DNA segment makes to the amino-acid composition of a polypeptide chain (Waters 2007). Thus, it may be said that while biochemists focus primarily on the "flow of matter or energy in the mechanism," molecular biologists trace the "flow of information" (Darden 2006a; 2006b, Ch. 3).

Despite the prevalence of functional talk in biology, it is surprisingly difficult to pinpoint the ontological status of functions. Some functional relationships highlighted in abstract models are explicitly instrumental, being understood as mere mathematical shortcuts, while others are laws from physics and chemistry presumably embodying some form of nomological necessity. In contrast, functional ascriptions related to claims about biological significance and biological information seem to be nothing but a linguistic device for describing what happens when interventions are conducted. When developmental biologists state that the function of the *wingless* gene is to determine wing development or body axis formation, they are describing differences between phenotypes of *Drosophila* with (test or mutated) and without mutations (wild-type or control) at the *wingless* locus. More complex examples refer to cases of functional equivalence, understood as invariance to intervention. Organic chemistry, biochemistry, and molecular biology routinely rely on functional classifications of molecular species. Such classifications are meant to capture the fact that substitutions of molecular species sharing certain features result in smaller differences to chemical and biological outcomes, as compared with differences resulting from substitutions by species that don't share these features. For instance, the exchange of leucine for the chemically similar isoleucine in the amino-acid sequence of a protein usually results in undetectable differences in geometry, binding affinity, or catalytic activity and for this reason is often treated as a silent mutation. In contrast, a point mutation in the beta-hemoglobin gene resulting in the replacement of a hydrophilic glutamic acid with a hydrophobic valine is the single most important genetic determinant of sickle-cell anemia. Other examples of functional equivalence include gene duplications, which are thought to explain why some gene knockout experiments are inconclusive, and the redundancy of the genetic code, which explains why certain genetic mutations make no difference to the amino-acid composition of proteins.

The so-called 'functional assays' commonly used in experimental biology don't seem to point to a more sophisticated interpretation either. Despite their name, these are simply tests for measuring something. Psychologists call this something a 'construct', biologists call it a 'function' or 'activity'. For instance, a luciferase reporter gene assay measures levels of gene expression. A TUNEL assay measures apoptosis. This doesn't tell us in what sense gene expression or apoptosis are 'functions'. Gene expression is DNA being transcribed and translated. The luciferase assay detects the amount of gene product (luciferase) produced. Apoptosis is cells digesting their DNA and proteins. A TUNEL assay detects the catalytic activity of caspase 3. It is difficult to see how a functional assay can reveal what a function is over and above a generic placeholder to be filled by whatever is measured.

In contrast to the above analysis, which accounts for functional ascriptions from the standpoint of experimental practice, Robert Cummins (1975) argues that functional roles should be analyzed from a theoretical perspective as capacities of parts that appear in an explanation of the capacity of a containing system. According to a Cummins-style analysis, prostaglandins function as ligands for receptors relative to an explanatory account of an organism's capacity to mount inflammatory responses following exposure to harmful stimuli inasmuch as prostaglandins are capable of binding receptors in an organism and the proposed explanation correctly accounts for the organism's capacity to mount inflammatory responses in part by appealing to the capacity of prostaglandins to bind receptors. Craver further identifies the explanatory account in Cummins' analysis as being a mechanistic explanation. Thus, the capacity of a part to play a functional role may be analyzed "in terms of the properties or activities by virtue of which it contributes to the working of a containing mechanism, and in terms of the mechanistic organization by which it makes that contribution" (2001, 61).

One shortcoming of Cummins' account is that it presupposes an already available explanation of a system's behavior. This requirement is at odds with the fact that, in scientific practice, functional ascriptions can be made before any specific explanation is hypothesized, let alone corroborated and accepted. Nevertheless, this contradiction can be solved by making the plausible proposal that functional ascriptions are revised, becoming more specific as an explanatory account emerges. For example, since mutations of the *wingless* gene in *Drosophila* resulted in abnormal wings, it was initially claimed that the function of the gene is to promote wing development. Once it was discovered that the gene is involved in the generation of a segmentation pattern of the Drosophila embryo, the function of the gene was revised accordingly: *wingless* contributes to wing development by playing a role in body axis formation. The revised functional ascription brings us closer to Cummins' account.

Whether one endorses an abstract model approach, an experimental analysis, or Cummins' theoretical account, functional ascriptions don't seem to introduce a novel physical interpretation. An experimental analysis reduces functional ascriptions to claims about biological significance, that is, to experimental results. Thus construed, functional ascriptions don't require or imply anything over and above the minimal experimental interpretation. In an abstract model or under Cummins' theoretical treatment, functional ascriptions presuppose an already existing interpretation of experimental results in terms of nomic necessity, capacities, structures, properties, interactions, and other items postulated by scientific explanations. It seems therefore reasonable to conclude that talk of biological functions, and perhaps information as well, neither presupposes nor entails a distinctive ontology but is rather a flexible mode of representation compatible with multiple interpretations.

4.7 Activities and Productive Causation

According to a highly influential characterization proposed by Peter Machamer, Lindley Darden, and Carl Craver, "mechanisms are entities and activities organized such that they are productive of regular changes from start-up or set-up to finish or termination conditions" (2000, 3). Phenomena are understood as 'regular changes from start-up or set-up to finish or termination conditions', a view compatible with the 'phenomena as reproducible data' account defended in Section 2. The remaining elements of the characterization, namely the notions of 'entity', 'activity', and 'organization', closely mirror the language scientists use to describe mechanisms. Activities, denoted by verbs, are seen as "the producers of change," while entities, denoted by nouns, are "the things that engage in activities" in virtue of "specific types of properties" (Machamer et al. 2000, 3). Organization refers to the spatial arrangement and intrinsic geometry of entities, and to the temporal order and duration of activities. For instance, prostaglandins bind receptors in virtue of a geometrical fit between the two molecules; then, binding brings about a conformational change in the geometry of the receptors, altering their chemical affinity for downstream protein mediators of various intracellular signaling pathways.

It has been argued that mechanistic descriptions found in the scientific literature are indicative of a dual ontology of entities and activities distinct from that of properties and interactions discussed earlier (Darden 2008; Machamer et al. 2000). The distinction is drawn on the grounds that activities are not reducible to properties or state transitions of parts and that, unlike interactions, they may involve only one part. Nevertheless, it is worth noting that both activities and interactions require parts having certain properties, in both cases parts are undergoing changes, and both fall into the same fundamental categories of "geometrico-mechanical, electro-chemical, electro-magnetic and energetic" (Machamer et al. 2000, 22) borrowed from physics and chemistry. As for divergences concerning the number of relata, they may be artefacts created by particular modes of linguistic representation. For instance, the spontaneous conformational change of a molecule is commonly described as an activity involving a single part, the molecule undergoing it. Yet, physics dictates that such descriptions can only be shorthands for a more complex story. Conformational changes require collisions with molecules present in the environment, such as those of the water solvent, which provide the energy necessary for the conformational change.[20]

Just like functions, activities are an assorted bunch: some have a physical interpretation, while others are nothing else but descriptions of experimental results. For instance, 'phosphorylation' and 'hydrolysis' refer to known chemical processes. In contrast, claims about 'activation/inhibition', 'upregulation/downregulation', 'expression/suppression' are nothing but qualitative data describing how the values of measured variables compare between test and control conditions. The fact that the terms 'function' and 'activity' are used to describe both data and physical structures, processes, and interactions suggests that they are umbrella terms indicative of a pluralist stance about physical interpretation and not genuine ontological categories. If this conclusion is granted, then we should exercise caution when attempting to extract ontology out of descriptions of mechanisms found in the scientific literature. More specifically, we should resist the temptation of assuming that differences in

---

[20] In a more conciliatory tone, James Tabery (2004, 9) concludes that the "requirement of productivity, rather than demanding an ontological switch from Glennan's interactions to activities, only reveals the need for interactions as Glennan conceives them alongside activities."

terminology and modes of representation are automatically indicative of differences in the physical nature of mechanisms or that similar forms of representation amenable to general characterizations reflect core characteristics of mechanisms.

The proposal of a dual 'entities and activities' ontology led some authors to further argue that mechanisms are associated with an 'active' or 'productive' notion of causation, as opposed to a 'passive' conception defining causation relative to what happens on other occasions (regularity) or in contrast to other situations (difference-making) (Bogen 2008; Illari and Williamson 2011; Machamer 2004). The classical argument for drawing a distinction between two notions of causation rests on an analysis of cases of causal overdetermination. Ned Hall (2004, 235) gives the following example: "Suzy and Billy, expert rock-throwers, are engaged in a competition to see who can shatter a target bottle first. They both pick up rocks and throw them at the bottle, but Suzy throws hers a split second before Billy. Consequently Suzy's rock gets there first, shattering the bottle. Since both throws are perfectly accurate, Billy's would have shattered the bottle if Suzy's had not occurred, so the shattering is overdetermined." The example illustrates the fact that a test for causal relevance gives a negative result when applied to failsafe mechanisms, such as ones relying on redundant causal pathways. Yet, irrespective of the test's result, we – the reader-subjects in this psycho-philosophical experiment – conclude that Suzy's throw caused the shattering of the bottle. This, Hall believes, shows that the result is a false negative and that we must rely on a second and more fundamental notion of causation, a productive one having something to do with the satisfaction of the desiderata of transitivity, locality (evidence for spatiotemporally continuous chains of causal intermediates connecting cause and effect), and intrinsicness (the notion that the causal nature of a situation is determined by its intrinsic properties).

Philosophical intuitions aside, Hall's experiment is inconclusive. The most pressing problem is that it fails to control for alternative explanations. The same conclusion that Suzy's throw caused the shattering of the bottle may be reached from the following premises: the causal relevance of collisions vis-à-vis the shattering of bottles; the accuracy of the throws, which guarantees collisions; the fact that the shattering of bottles is an irreversible process (i.e., a bottle cannot be shattered twice); and the fact that Suzy's rock collides with the bottle first. It is true, the inference cannot be made based on knowledge of difference-making dependencies alone. It requires additional information. This information, however, has nothing to do with a second notion of causation. It concerns only contrasts, regularities, and a particular about the case.[21] A second problem is that the experiment doesn't specifically test the hypothesis that the second notion of causation involves productivity.

Experiments in empirical psychology control for prior knowledge of causal relevance while providing cues about spatiotemporal continuity by presenting subjects with visual displays consisting of moving geometric shapes (Scholl and Tremoulet 2000). The results vindicate Hall's intuition: children readily identify certain spatiotemporal patterns, such as those mimicking a collision or a chase, as causal relationships indicative of mechanical or social interactions. It's a mitigated victory though. This type of causal judgment is thought to rely on an automatic form of visual processing responding to a narrow range of spatiotemporal patterns. It can therefore account for only a small fraction of causal judgments. Another limitation is that judgments relying on locality considerations are prone to false positives, and in this sense constitute a type of perceptual illusion. For instance, if shown a video depicting a spatiotemporally continuous sequence of causal intermediates beginning with Suzy throwing a ball and ending with a car exploding when hit by the ball, would we conclude that the cause of the explosion is Suzy's throw? The fact that, unlike children, adult subjects resist the conclusion suggests that the desiderata of causal productivity outlined by Hall play a role in causal judgments made in the absence of prior experience, but are trumped by causal learning, associative or inferential, relying on a difference-making notion of causation. These limitations undermine the credibility of the proposal that causal productivity is more fundamental, more general, and ultimately likely to play a decisive role in the physical interpretation of experimental results.

Perhaps Stuart Glennan's recent defense of a singularist, productive, and mechanist account of causation captures most vividly two shortcomings of the passive conceptions of causation underpinning the minimal experimental interpretation. The first is that it is highly unlikely that there is such a thing as a single feature of reality corresponding to the notion of 'causal relationship'. Rather, Glennan argues, "the utility of the word 'cause' is that it allows us to assert

---

[21] An analogous example from science is gene duplication. The knockout of one or the other of the copies has no effect on the phenotype. This seems to indicate that neither copy is causally relevant. A double knockout, however, turns out to have marked consequences for the phenotype. Collectively, the three experiments indicate that the two genes are part of a mechanism relying on redundant causal pathways. Now, given that it is not known how these genes come to have an effect on the phenotype, and therefore it is not possible to detail a continuous chain of causal intermediates or to identify intrinsic characteristics indicative of causation, can it be established which of the copies is responsible for an outcome in a given situation, say, physiological ('normal') conditions? As a rule, yes. The simplest strategy is to check gene expression levels under physiological conditions. If, by default, one copy is expressed and the other silenced, it can be concluded that only the expressed copy contributes to the phenotype.

causal dependence without saying anything about the kind of activities and interactions that ground the dependence" (2017, 156). In other words, talk of causes captures a multitude of distinct physical processes satisfying a common set of desiderata, such as the methodological requirements of a controlled experiment outlined in Section 3.1.

Indeed, there is no reason to assume that a homogeneous class of objects satisfy generic desiderata. Nevertheless, it is difficult to see how the notion of causal productivity solves the problem. If causal productivity is understood as the satisfaction of a different set of desiderata, such as Hall's requirements of locality and intrinsicness, then it too is a generic concept. It doesn't tell us which specific spatiotemporal patterns and characteristics are involved in a given instance of causal productivity. On the other hand, if causal productivity is understood as the claim that reality consists of singular or strictly local causal processes, as suggested by several authors (Anscombe 1993; Bogen 2004; 2005; Darden 2008; Glennan 2011; 2017; Illari and Williamson 2011; Menzies 1996), then this claim cannot be justified empirically. Causal singularities cannot be discovered by means of empirical inquiry, as there is no experimental methodology capable of disentangling chance associations from instances of singular causation or localizing an instance of singular causation within the boundaries of a particular physical system (Baetu 2013). A theoretical justification is also difficult to mount. Even if rare, or perhaps even singular, behaviors of systems may be predicted given prior knowledge of how the parts of the system interact, this still requires a theory of the parts and their interactions, which in turn embodies prior knowledge of what happens on other occasions. From a scientific point of view, the only defensible strategy is a middle path: acknowledge that causal relevance claims may have more than one physical interpretation but abstain from jumping to the conclusion that reality consists of singular causal structures. I think this is precisely what Machamer, Darden, and Craver (Darden 2008, 962–63; Machamer et al. 2000, 22) are doing when they are classifying mechanistic activities as mechanical, thermodynamic, and chemical processes.

The second point of dissatisfaction raised by Glennan is that passive conceptions are analogous to operationalized definitions, equating causal relationships to epistemic methods for demonstrating causation without ever telling us what features of reality are captured by these methods (2017, 146). The expectation here is that a scientific explanation should provide a physical interpretation over and above the minimal experimental interpretation.

I agree. The fact that at least some causal dependencies appearing in mechanistic explanations have a physical interpretation justifies the claim that something more substantive can and should be said about the nature of causal relationships. Furthermore, mechanistic explanations are expected to be complete, in the sense that they describe causal structures sufficient to produce a phenomenon. Yet webs of difference-making causal dependencies don't provide enough information about causal sufficiency. That an intervention on determinant X results in a change in outcome Y demonstrates causal relevance, but does not prove that X is sufficient to produce Y, or that X is the only determinant of Y. If this is true, then aggregating causally relevant factors into ever more elaborate causal webs cannot get us any closer to proving causal sufficiency either. If anything, causal webs are bound to remain open-ended, as new causal factors (e.g., new drugs and technologies allowing for novel experimental interventions) can always be appended to the web. It would seem therefore that explanatory completeness requires a shift from statements of causal relevance indicating that a phenomenon is susceptible to manipulation via interventions on individual mechanistic components, to the bolder claim that the mechanism as a whole is sufficient to produce and is actually producing a phenomenon in a given experimental or biological context (Baetu 2016b; Fagan 2012). Glennan argues that the missing ingredient is a metaphysical notion of causal productiveness. In contrast, I think we should rely on what science tells us about the nature of physical reality, which is not a story about productive and singular causation. Thus far, what scientific inquiry revealed about reality is that it is far stranger and more complex than any prêt-à-porter metaphysical interpretations, hence the variegated assortment of ontological categories presented in this section.

4.8 An Overview and Classification of Mechanistic Accounts

The above survey reveals that mechanistic ontologies fall into three main categories. Considerations pertaining to part–whole composition, interactions, identity, and structure add novel elements to the minimal experimental interpretation, extending it into the full-blown mechanistic ontology representative of most research in cell and molecular biology. In particular, the localization of variables targeted by measurements and interventions justifies the claim that the causal structures responsible for phenomena consist of organized systems of parts, while theoretical considerations from physics and chemistry allow for an interpretation of some causal dependencies between variables as physical interactions between parts. The ·notion that mechanisms are, ontologically speaking, organized systems of interacting parts doesn't please everybody. This conception of mechanisms is sometimes shunned by philosophers due to its reductionistic undertones and its tacit reliance on the notion of 'law of nature', as well as by methodologists, clinicians, and scientists engaged in interdisciplinary research, who tend to prefer the more neutral minimal experimental interpretation. Nevertheless, despite objections raised by dissatisfied parties, a 'parts and interactions' type of ontology remains the cornerstone of most physical interpretations espoused in scientific circles ("A Mechanistic Meeting Point" 2007).

Analyses of linguistic descriptions of mechanisms, as typically documented in science textbooks, support a different class of characterizations in terms of functions of parts and of a dual ontology of entities and activities. It is not clear to what extent these characterizations introduce novel ontologies. Although most authors insist in treating activities and functions as fundamental ontological items, in practice, talk of functions and activities needs to be disambiguated on a case-by-case basis, as it may refer to physical interactions, processes, or structures, theoretical constructs postulated by putative or accepted explanations, or may simply amount to linguistic devices for summarizing or reporting experimental results. I think therefore that the main strength of these characterizations lies not in their intended ontological contributions, but rather in their ability to foster a pluralist stance with respect to physical interpretation. A pluralist stance has the notable advantage of providing a means for accommodating ontological complexity by introducing placeholder terms, such as 'function' and 'activity', where individual instances of the placeholder can be filled by whatever turns out to correspond to it in terms of physical reality. For as long as the placeholders themselves are not reified, the presupposition of a monolithic mechanistic ontology is avoided, allowing for the true nature of biological mechanisms to be discovered and reconstructed from bits and pieces of interpreted experimental results.

Finally, some ambitious metaphysical claims have been made concerning singular and productive mechanistic causation. According to mechanistic accounts of causation, mechanisms are not grounded in, or made possible by, causal dependencies, but rather mechanisms ground causal dependencies. The notion that some causal regularities depend on mechanisms is certainly not foreign to biology. For instance, the law-like regularity 'the multiplication of thoracic segments in tetrapods causes a loss of forelimbs' is underpinned by a molecular mechanism that makes it such that the expression of Hox-C6 in cervical segments, which is required for their transformation in thoracic segments, also results in an inhibition of the sonic hedgehog signaling pathway needed for limb specification (Cohn and Tickle 1999). One may therefore legitimately argue that at least some causal regularities depend in a strong ontological sense on mechanisms and that causation, especially alleged instances of singular causation inaccessible to standard experimental methodology, can be inferred based on prior knowledge of mechanisms. It remains to be seen to what extent a mechanistic approach to causation is generalizable. One worry that immediately comes to mind is that some causal dependencies refer to fundamental forces and modes of interaction. A second worry, which provided much of the impetus behind the evidence-based medicine movement, is that many causal inferences based on mechanistic rationales proved to be wrong. The worry I myself expressed in this Element is that proponents of mechanistic accounts of causation don't offer an alternative experimental methodology that would allow for a characterization of phenomena without appealing to the notion of reproducibility and for the discovery of mechanisms without relying on the difference-making notion of causation at the heart of controlled experiments.

A Recapitulation and Some Clarifications

The general line of argumentation developed in this Element can be summarized as follows. I initiated my inquiry under the guiding assumption that the significant contribution of experimental research to the elucidation of biological mechanisms – especially with the advent of molecular biology, which is responsible for some of the most famous mechanistic explanations in biology – justifies the demand that mechanistic metaphysics should take into account the metaphysical assumptions underlying the standard methodology governing most experimental research in the life sciences. Thus, I surmised, two kinds of considerations should constrain mechanistic metaphysics: a set of more fundamental, methodological considerations provide the building blocks for a minimal experimental interpretation, which in turn serves as a bare-bones infrastructure onto which more complex physical interpretations may be subsequently grafted.

I set up the task of investigating what basic methodological assumptions entail for a metaphysical conception of mechanisms by tackling a question seldom addressed nowadays: what exactly is a phenomenon? The standard answer is surprisingly elusive: phenomena are not data. Rather, we are told, phenomena have more to do with statistical patterns in data, being in fact a sort of theoretical construct. Before endorsing this result, I thought it would be useful to look closer into what phenomena are allegedly not. What are data, then? An analysis of basic procedures falling under the heading 'good experimental practice' led me to conclude that data are not just measured variable values, but rather sets of variable values obtained in such a way as to allow for the localization of the causes responsible for differences in measured values. By following this thread, I came to articulate an experimental characterization of phenomena as data reproduced when experiments are replicated. On this account, the successful characterization of a phenomenon has less to do with the identification of statistical patterns, and more with the ability to circumscribe a set of proximal causes within the spatiotemporal boundaries of physical systems satisfying a given description.

This is not to say that statistical analysis and inference are irrelevant. The informal patterns of reasoning associated with contrastive inferences, such as Mill's method of difference, or those driving experimental fiddling aiming to enhance the reproducibility of data, have mathematically rigorous parallels in modern statistics. By itself, however, mathematical rigor doesn't tell us how patterns of inference relate to physical reality, which is what I was after in order to avoid the constructivist implications of the view that phenomena are nothing but the outputs of a statistical processing of measurements. For instance, there is a similarity between the account of phenomena proposed in this Element and latent variable models in psychometric research, such as Spearman's (1904) famous model of general intelligence. The general pattern of reasoning is similar in the sense that the existence of a hidden causal structure is inferred based on evidence for a set of correlations between measured variables. The difference lies in the fact that the account I defend hinges crucially on experimental intervention – on how physical changes in an experimental setup result in gains or losses in reproducibility. I argued that the fact that reproducibility can be enhanced by modifying experimental setups demonstrates that data are not categorically 'clean' or 'noisy', reproducible or irreproducible, but rather come in degrees of noisiness and reproducibility, depending on the experimental setup used to generate them. In contrast, factor analysis doesn't involve any fiddling with the experimental setup but relies strictly on a statistical analysis of available variable values.

Armed with an experimental characterization of phenomena, I proceeded to the next major issue a metaphysical account should address, namely the nature of the relationship between mechanisms and phenomena. The fact that typical characterizations of phenomena in biological and clinical science embody substantive knowledge about the causal structure of the world and that, moreover, the elucidation of mechanisms invariably relies on experiments designed to demonstrate causal relevance provided compelling grounds to believe that mechanisms cause phenomena. The leading causal account in the philosophical literature, however – the etiological account – takes phenomena to be time-point changes in the values of measured variables. Thus construed, phenomena are not informative about their causes. I insisted therefore that an adequate characterization of a phenomenon requires the specification of an experimental setup fixing the circumstances in which the value of a variable is shown to change. For instance, only 'chorea induced by damage to the cerebellum' or 'autosomal-dominant inherited chorea' are genuine phenomena. In contrast, some historically documented instance of chorea is not a phenomenon whose causes can be elucidated by experimental methods.

Still, some may retort that epidemiologists are routinely engaged in the business of discovering risk factors associated with or causally relevant to a given outcome, such as chorea. True. There is, however, an important difference between the epidemiologist's statistical notion of 'etiology in a population' and the philosopher's notion of 'historical etiology'. Epidemiologists do specify experimental setups, namely the populations in which they document the incidence of factors and outcomes. In other words, what epidemiologists investigate is not the cause of some particular instance of chorea, but the many and diverse causal factors responsible for the prevalence of chorea in a clearly defined population. Moreover, the populations initially studied by epidemiologists are eventually narrowed down in subsequent basic and clinical research until, as illustrated in the case of Huntington's disease, an experimental model of a phenomenon emerges.

The alternative, and much more popular, constitutive account explicitly denies that mechanisms cause phenomena. The key intuition here is that mechanisms are organized systems of parts. This is naturally conducive to the notion that parts must be relevant to whatever the system is doing, which is usually the phenomenon to be explained, while the organization of the system constrains or channels the various activities performed by the parts. In short, the parts must be relevant to the behavior of the whole, and the whole must be relevant to the behavior of the parts. But since parts and wholes are not spatiotemporally distinct in the way we expect causes and effects to be, mechanisms cannot cause phenomena. In response to this proposal, I argued that the part–whole constitutive interpretation fails to map congruently on any known experimental design. Measurements and experimental interventions target variables – or the factors, properties, aspects, or features of reality to which these variables correspond – not parts and wholes. One can never measure a table simpliciter, but only the length, the weight, or the spectral reflectance of a table. To be sure, one can remove a leg from the table and see what happens, but this doesn't eliminate the methodological requirement of a proper characterization of the variables and variable values hinted at by expressions such as 'remove table leg' and 'see what happens'. At the very least, one must specify the techniques of measurement and intervention involved.

My own attempt to develop a causal account of the relationship between mechanisms and phenomena reflects a double commitment to a causal account of measurement and to the view that phenomena are data informative of the causal structure of the world. What I called the 'causal mediation account' is a generalization of the notion of causal mediation, as demonstrated by a common type of experiment deployed in mechanistic research: the knock-out experiment. On this account, a phenomenon is a measured stimulus–response sequence. The mechanism responsible for a phenomenon is a mediating causal structure linking, via a certain number of causal intermediaries, the factors targeted by the measurements involved in the characterization of the phenomenon. The mechanism, which is to say the presence of a mediating causal structure in an experimental setup, causally ensures the stability, and therefore the reproducibility, of the phenomenon in that experimental setup. I say 'causally' because disruptions of the mediating mechanism – as, for instance, those involved in knock-out experiments – cause, in a difference-making sense, changes in the reproducibility of the phenomenon. Relative to the etiological account, the advantage of this approach is that a bona fide phenomenon is already assumed, namely a consistently reproducible stimulus–response association in a specified experimental setup. Compared to the constitutive account, there are parts and wholes, but they are now understood as causal chains being part of other causal chains, so that compatibility with standard experimental methodology is preserved.

The last question on my agenda was, What else can be said about the physical nature of mechanisms? A great deal more, it turns out, to the point that I cannot claim to have done full justice to the multitude of physical interpretations and mechanistic characterizations available in the scientific and philosophical literatures. The most pressing issue was to rehabilitate one of the key intuitions behind the constitutive account, namely that mechanisms are systems of interacting parts. I discussed how this may be achieved by combining the notions of localization and decomposition/recomposition with a theoretical background from physics and chemistry. Toward the end of the Element, I discussed recent claims about dualistic entities/parts and activities/functions ontologies, as well as the notion that mechanisms may be ontologically more fundamental than causation. My general assessment of these proposals is deflationary. By 'deflationary' I do not simply mean to dismiss these novel theses as impractical or overly speculative. Rather, my point of view is that although they do reflect something important about how scientists think about mechanisms, more work needs to be done in order to zero in on what exactly this something may be.

References

Ankeny, R. 2001. "Model Organisms as Models: Understanding the 'Lingua Franca' of the Human Genome Project." *Philosophy of Science* 68:S251–S61.

Ankeny, R., and S. Leonelli. 2011. "What's So Special about Model Organisms?" *Studies in History and Philosophy of Science* 42:313–23.

Anscombe, G. E. M. 1993. "Causality and Determination." In *Causation*, ed. E. Sosa and M. Tooley. Oxford: Oxford University Press.

Asmundson, G. J. G., and K. D. Wright. 2004. "Biopsychosocial Approaches to Pain." In *Pain: Psychological Perspectives*, ed. T. Hadjistavropoulos and K. D. Craig. Mahwah, NJ: Lawrence Erlbaum Associates.

Baetu, T. M. 2012a. "Filling In the Mechanistic Details: Two-Variable Experiments as Tests for Constitutive Relevance." *European Journal for Philosophy of Science* 2 (3):337–53.

2012b. "Mechanistic Constraints on Evolutionary Outcomes." *Philosophy of Science* 79 (2):276–94.

2013. "Chance, Experimental Reproducibility, and Mechanistic Regularity." *International Studies in History and Philosophy of Science* 27 (3):255–73.

2014. "Models and the Mosaic of Scientific Knowledge. The Case of Immunology." *Studies in History and Philosophy of Biological and Biomedical Sciences* 45:49–56.

2015a. "From Mechanisms to Mathematical Models and Back to Mechanisms: Quantitative Mechanistic Explanations." In *Explanation in Biology. An Enquiry into the Diversity of Explanatory Patterns in the Life Sciences*, ed. P.-A. Braillard and C. Malaterre, 345–63. Dordrecht: Springer.

2015b. "When Is a Mechanistic Explanation Satisfactory? Reductionism and Antireductionism in the Context of Mechanistic Explanations." In *Romanian Studies in the History and Philosophy of Science*, ed. G. Sandu, I. Parvu and I. Toader. Dordrecht: Springer.

2016a. "The 'Big Picture': The Problem of Extrapolation in Basic Research." *British Journal for the Philosophy of Science* 67 (4):941–64.

2016b. "From Interventions to Mechanistic Explanations." *Synthese* 193:3311–27.

2017a. "Mechanisms in Molecular Biology." In *Routledge Handbook of Mechanisms*, ed. Stuart Glennan and Phyllis Illari. New York: Routledge.

2017b. "On Pain Experience, Interdisciplinary Integration and Levels of Description, Explanation and Reality." *Synthese* DOI 10.1007/s11229-017-1429-5.

2017c. "On the Possibility of Designing Crucial Experiments in Biology." *British Journal for the Philosophy of Science.*

Baetu, T. M. 2019. "On the Possibility of Designing Crucial Experiments in Biology." *British Journal for the Philosophy of Science* 70 (2):407–429.

Ball, P. 2011. "Physics of Life: The Dawn of Quantum Biology." *Nature* 474 (7351):272–74.

Barwich, A.-S. 2015. "Bending Molecules or Bending the Rules: The Application of Theoretical Models in Fragrance Chemistry." *Perspectives on Science* 23 (4):1–23.

Bates, G. P. 2005. "The Molecular Genetics of Huntington Disease – A History." *Nature Reviews Genetics* 6:766–73.

Baumgartner, M., and A. Gebharter. 2016. "Constitutive Relevance, Mutual Manipulability, and Fat-Handedness." *British Journal for the Philosophy of Science* 67:731–56.

Bechtel, W. 2006. *Discovering Cell Mechanisms: The Creation of Modern Cell Biology.* Cambridge: Cambridge University Press.

2008. *Mental Mechanisms: Philosophical Perspectives on Cognitive Neuroscience.* New York: Routledge.

2009. "Generalization and Discovery by Assuming Conserved Mechanisms: Cross Species Research on Circadian Oscillators." *Philosophy of Science* 76:762–73.

2011. "Mechanism and Biological Explanation." *Philosophy of Science* 78:533–57.

Bechtel, W., and A. Abrahamsen. 2005. "Explanation: A Mechanist Alternative." *Studies in History and Philosophy of Biological and Biomedical Sciences* 36:421–41.

2010. "Dynamic Mechanistic Explanation: Computational Modeling of Circadian Rhythms as an Exemplar for Cognitive Science." *Studies in History and Philosophy of Science Part A* 41:321–33.

Bechtel, W., and R. Richardson. 2010. *Discovering Complexity: Decomposition and Localization as Strategies in Scientific Research.* Cambridge: Massachusetts Institute of Technology Press.

Bickle, J. 2006. "Reducing Mind to Molecular Pathways: Explicating the Reductionism Implicit in Current Cellular and Molecular Neuroscience." *Synthese* 151:411–34.

Bogen, J. 2004. "Analyzing Causality: The Opposite of Counterfactual is Factual." *International Studies in the Philosophy of Science* 18:3–26.

2005. "Regularities and Causality; Generalizations and Causal Explanations." *Studies in History and Philosophy of Biological and Biomedical Sciences* 36:397–420.

2008. "Causally Productive Activities." *Studies in History and Philosophy of Science* 39:112–23.

Bogen, J., and J. Woodward. 1988. "Saving the Phenomena." *The Philosophical Review* 97 (3):303–52.

Botting, R. M. 2010. "Vane's Discovery of the Mechanism of Action of Aspirin Changed Our Understanding of Its Clinical Pharmacology." *Pharmacological Reports* 62:518–25.

Braillard, P.-A. 2014. "Prospect and Limits of Explaining Biological Systems in Engineering Terms." In *Explanation in Biology. An Enquiry into the Diversity of Explanatory Patterns in the Life Sciences*, ed. P.-A. Braillard and C. Malaterre. Dordrecht: Springer.

Bridgman, P. 1927. *The Logic of Modern Physics*. New York: Macmillan.

Broadbent, A. 2013. *Philosophy of Epidemiology*. Houndmills: Palgrave Macmillan.

Brooks, D. S. 2017. "In Defense of Levels: Layer Cakes and Guilt by Association." *Biological Theory* 12 (3):142–56.

Brown, J. R. 1994. *Smoke and Mirrors: How Science Reflects Reality*. London: Routledge.

Brune, K., and B. Hinz. 2004. "The Discovery and Development of Antiinflammatory Drugs." *Arthritis & Rheumatism* 50 (8):2391–99.

Burnham, K. P., and D. R. Anderson. 2002. *Model Selection and Multimodel Inference: A Practical Information-Theoretic Approach*. 2nd ed. New York: Springer.

Cartwright, N. 1989. *Nature's Capacities and Their Measurement*. Oxford: Oxford University Press.

Cohn, M., and C. Tickle. 1999. "Developmental Basis of Limblessness and Axial Patterning in Snakes." *Nature* 399:474–79.

Craver, C. 2001. "Role Functions, Mechanisms, and Hierarchy." *Philosophy of Science* 68:53–74.

2007. *Explaining the Brain: Mechanisms and the Mosaic Unity of Neuroscience*. Oxford: Clarendon Press.

Craver, C., and W. Bechtel. 2007. "Top-Down Causation without Top-Down Causes." *Biology and Philosophy* 22:547–63.

Craver, C., and L. Darden. 2013. *In Search of Biological Mechanisms: Discoveries across the Life Sciences*. Chicago: University of Chicago Press.

Cremer, T., and C. Cremer. 2001. "Chromosome Territories, Nuclear Architecture and Gene Regulation in Mammalian Cells." *Nature Reviews Genetics* 2:292–301.

Cummins, R. 1975. "Functional Analysis." *Journal of Philosophy* 72 (20):741–65.

2000. "'How Does It Work' Versus 'What Are the Laws?': Two Conceptions of Psychological Explanation." In *Explanation and Cognition*, ed. F. Keil and R. Wilson, 117–45. Cambridge: Massachusetts Institute of Technology Press.

Darden, L. 1991. *Theory Change in Science: Strategies from Mendelian Genetics*. New York: Oxford University Press.

2006a. "Flow of Information in Molecular Biological Mechanisms." *Biological Theory* 1 (3):280–87.

2006b. *Reasoning in Biological Discoveries: Essays on Mechanisms, Interfield Relations, and Anomaly Resolution*. Cambridge: Cambridge University Press.

2008. "Thinking Again about Biological Mechanisms." *Philosophy of Science* 75:958–69.

Davidson, E., and M. Levine. 2005. "Gene Regulatory Networks." *Proceedings of the National Academy of Science* 102 (14):4935.

Dowe, P. 1995. "Causality and Conserved Quantities: A Reply to Salmon." *Philosophy of Science* 62:321–33.

Ellis, J. 2001. "Macromolecular Crowding: Obvious but Underappreciated." *Trends in Biochemical Sciences* 26 (10):597–604.

Elowitz, M., A. J. Levine, E. D. Siggia, and P. S. Swain. 2002. "Stochastic Gene Expression in a Single Cell." *Science* 297 (5584):1183–86.

Fagan, M. B. 2012. "The Joint Account of Mechanistic Explanation." *Philosophy of Science* 79 (4):448–72.

Fisher, R. A. 1935. *The Design of Experiments*. Edinburgh: Oliver and Boyd.

Frith, C. D. 1992. *The Cognitive Neuropsychology of Schizophrenia*. Hove: Lawrence Erlbaum Associates.

Germain, P.-L., and T. M. Baetu. 2017. "Extrapolation in Biomedical Research: A Multi-Model Perspective." In *Foundational Issues in Molecular Medicine*, ed. Marco Nathan and Giovanni Boniolo. New York: Routledge.

Glennan, S. 1996. "Mechanisms and the Nature of Causation." *Erkenntnis* 44:49–71.

2002. "Rethinking Mechanistic Explanation." *Philosophy of Science* 69: S342–S53.

2010. "Ephemeral Mechanisms and Historical Explanation." *Erkenntnis* 72:251–66.

2011. "Singular and General Causal Relations: A Mechanist Perspective." In *Causality in the Sciences*, ed. P. McKay, J. Williamson and F. Russo, 789–817. Oxford: Oxford University Press.

2017. *The New Mechanical Philosophy*. New York: Oxford University Press.

Goodsell, D. S. 2009. *The Machinery of Life*. 2nd ed. New York: Copernicus Books.

Gori, G. B. 1989. "Epidemiology and the Concept of Causation in Multifactorial Diseases." *Regulatory Toxicology and Pharmacology* 9 (3):263–72.

Gross, F. 2015. "The Relevance of Irrelevance: Explanation in Systems Biology." In *Explanation in Biology. An Enquiry into the Diversity of Explanatory Patterns in the Life Sciences*, ed. P.-A. Braillard and C. Malaterre. Dordrecht: Springer.

Hacking, I. 1983. *Representing and Intervening*. Cambridge: Cambridge University Press.

1990. *The Taming of Chance*. Cambridge: Cambridge University Press.

Hall, N. 2004. "Two Concepts of Causation." In *Causation and Counterfactuals*, ed. J. Collins, N. Hall, and L. Paul, 225–76. Cambridge, MA: Massachusetts Institute of Technology Press.

Harinen, T. 2014. "Mutual Manipulability and Causal Inbetweenness." *Synthese* 195 (1):34–54.

Hendry, R. F. 2008. "Two Conceptions of the Chemical Bond." *Philosophy of Science* 75 (5):909–20.

Hill, A. B. 1955. *Principles of Medical Statistics*. 6th ed. New York: Oxford University Press.

1965. "The Environment and Disease: Association or Causation?" *Proceedings of the Royal Society of Medicine* 58:295–300.

Hochachka, P. W. 1999. "The Metabolic Implications of Intracellular Circulation." *Proceedings of the National Academy of Sciences* 96 (22):12233–39.

Hoffmann, A., A. Levchenko, M. Scott, and D. Baltimore. 2002. "The IκB–NF-κB Signaling Module: Temporal Control and Selective Gene Activation." *Science* 298:1241–45.

Hoffmann, P. M. 2012. *Life's Ratchet: How Molecular Machines Extract Order from Chaos*. New York: Basic Books.

Hohwy, J. 2007. "The Search for Neural Correlates of Consciousness." *Philosophy Compass* 2 (3):461–74.

Holmes, F. L. 2001. *Meselson, Stahl, and the Replication of DNA: A History of the Most Beautiful Experiment in Biology*. New Haven: Yale University Press.

2006. *Reconceiving the Gene: Seymour Benzer's Adventures in Phage Genetic*. New Haven: Yale University Press.

Howick, J. 2011. *The Philosophy of Evidence-Based Medicine*. Oxford: BMJ Books.

Illari, P., and J. Williamson. 2011. "Mechanisms Are Real and Local." In *Causality in the Sciences*, ed. P. McKay Illari, F. Russo, and J. Williamson, 818–44. Oxford: Oxford University Press.

2012. "What Is a Mechanism? Thinking about Mechanisms *across* the Sciences." *European Journal for Philosophy of Science* 2 (1):119–35.

Issad, T., and C. Malaterre. 2015. "Mechanisms, Models and Explanatory Force." In *Explanation in Biology. An Enquiry into the Diversity of Explanatory Patterns in the Life Sciences*, ed. P.-A. Braillard and C. Malaterre. Dordrecht: Springer.

Jocher, A., S. Kessler, S. Hornstein, J. Schulte Mönting, and C. M. Schempp. 2005. "The UV Erythema Test as a Model to Investigate the Anti-Infl ammatory Potency of Topical Preparations – Reevaluation and Optimization of the Method." *Skin Pharmacology and Physiology* 18:234–40.

Kaiser, M. I., and B. Krickel. 2017. "The Metaphysics of Constitutive Mechanistic Phenomena." *British Journal for the Philosophy of Science* 68 (3):745–79.

Kendall, R. E. 2001. "The Distinction between Mental and Physical Illness." *The British Journal of Psychiatry* 178 (6):490–93.

Kendler, K. S., and J. Campbell. 2009. "Interventionist Causal Models in Psychiatry: Repositioning the Mind-Body Problem." *Psychological Medicine* 39:881–87.

Kim, J. 2005. *Physicalism or Something Near Enough* Princeton, NJ: Princeton University Press.

Koch, C., M. Massimini, M. Boly, and G. Tononi. 2016. "Neural Correlates of Consciousness: Progress and Problems." *Nature Reviews Neuroscience* 17 (5):307–21.

Krieger, N. 1994. "Epidemiology and the Web of Causation: Has Anyone Seen the Spider?" *Social Science & Medicine* 39 (7):887–903.

Kupiec, J.-J., O. Gandrillon, M. Morange, and M. Silberstein, eds. 2011. *Le hasard au coeur de la cellule: Probabilités, déterminisme, génétique*. Paris: Éditions Matériologiques.

Langenbach, R., C. D. Loftin, C. Lee, and H. Tiano. 1999. "Cyclooxygenase-Deficient Mice. A Summary of Their Characteristics and Susceptibilities to Inflammation and Carcinogenesis." *Annals of the New York Academy of Sciences* 889:52–61.

Leighton, J. P. 2010. "Internal Validity." In *Encyclopedia of Research Design*, ed. N. J. Salkind. Thousand Oaks, CA: Sage.

Leuridan, B. 2012. "Three Problems for the Mutual Manipulability Account of Constitutive Relevance in Mechanisms." *British Journal for the Philosophy of Science* 63:399–427.

Love, A. 2012. "Hierarchy, Causation and Explanation: Ubiquity, Locality and Pluralism." *Interface Focus* 2:115–25.

Machamer, P. 2004. "Activities and Causation: The Metaphysics and Epistemology of Mechanisms." *International Studies in the Philosophy of Science* 18 (1):27–39.

Machamer, P., L. Darden, and C. Craver. 2000. "Thinking About Mechanisms." *Philosophy of Science* 67:1–25.

Mathews, C. K. 1993. "The Cell-Bag of Enzymes or Network of Channels?" *Journal of Bacteriology* 175 (20):6377–81.

McAllister, J. W. 1997. "Phenomena and Patterns in Data Sets." *Erkenntnis* 47:217–28.

2007. "A Mechanistic Meeting Point." *Nature Chemical Biology* 3 (3):127.

Menzies, P. 1996. "Probabilistic Causation and the Pre-Emption Problem." *Mind* 105 (417):85–117.

Mill, J. S. 1843. *A System of Logic, Ratiocinative and Inductive*. London: John W. Parker.

Morange, M. 2002. "The Gene: Between Holism and Generalism." In *Promises and Limits of Reductionism in the Biomedical Sciences*, ed. M. Van Regenmortel and D. Hull. Chichester: John Wiley & Sons.

2009. "Synthetic Biology: A Bridge between Functional and Evolutionary Biology." *Biological Theory* 4 (4):368–77.

Morris, R. G. M. 1981. " Spatial Localization Does Not Require the Presence of Local Cues." *Learning and Motivation* 12 (2):239–60.

Murphy, D. 2006. *Psychiatry in the Scientific Image*. Cambridge, MA: Massachusetts Institute of Technology Press.

Netea, M. G., F. Balkwill, M. Chonchol, F. Cominelli, M. Y. Donath, E. J. Giamarellos-Bourboulis, D. Golenbock, M. S. Gresnigt, M. T. Heneka, H. M. Hoffman, R. Hotchkiss, L. A. B. Joosten, D. L Kastner, M. Korte, E. Latz, P. Libby, T. Mandrup-Poulsen, A. Mantovani, K. H. G. Mills, K. L. Nowak, L. A. O'Neill, P. Pickkers, T. van der Poll, P. M. Ridker, J. Schalkwijk, D. A. Schwartz, B. Siegmund, C. J. Steer, H. Tilg, J. W. M. van der Meer, F. L. van de Veerdonk, and C. A. Dinarello. 2017. "A Guiding Map for Inflammation." *Nature Immunology* 18:826–31.

Philips, R., and S. R. Quake. 2006. "The Biological Frontier of Physics." *Physics Today* 59: 38–43.

Potochnik, A., and B. McGill. 2012. "The Limitations of Hierarchical Organization." *Philosophy of Science* 79 (1):120–40.

Ptashne, M. A. 1998. *Genetic Switch: Phage Lambda and Higher Organisms*. Cambridge, MA: Blackwell Scientific Publications.

Rainsford, K. 2015. "History and Development of Ibuprofen." In *Ibuprofen: Discovery, Development and Therapeutics*, ed. K. Rainsford, 1–21. Chichester: John Wiley & Sons.

Rao, C. V., D. M. Wolf, and A. P. Arkin. 2002. "Control, Exploitation and Tolerance of Intracellular Noise." *Nature* 420 (6912):231–37.

Romero, F. 2015. "Why There Isn't Inter-Level Causation in Mechanisms." *Synthese* 192:3731–55.

Rosenberg, A. 2006. *Darwinian Reductionism, or, How to Stop Worrying and Love Molecular Biology*. Chicago: University of Chicago Press.

Salmon, W. 1984. *Scientific Explanation and the Causal Structure of the World*. Princeton: Princeton University Press.

1997. "Causality and Explanation: A Reply to Two Critiques." *Philosophy of Science* 64:461–77.

Schaffner, K. F. 1993. *Discovery and Explanation in Biology and Medicine*. Chicago: University of Chicago Press.

Scholl, B. J., and P. D. Tremoulet. 2000. "Perceptual Causality and Animacy." *Trends in Cognitive Sciences* 4 (8):299–309.

Schwarz, H. P., and F. Dorner. 2003. "Karl Landsteiner and His Major Contributions to Haematology." *British Journal of Haematology* 121 (4):556–65.

Smart, J. J. C. 1959. "Sensations and Brain Processes." *Philosophical Review* 68:141–56.

Smyth, J. M., M. C. Ockenfels, A. A. Gorin, D. Catley, L. S. Porter, C. Kirschbaum, D. H. Hellhammer, and A. A. Stone. 1997. "Individual Differences in the Diurnal Cycle of Cortisol." *Psychoneuroendocrinology* 22 (2):89–105.

Spearman, C. E. 1904. "'General Intelligence', Objectively Determined and Measured." *American Journal of Psychology* 15:201–93.

Steel, D. 2007. *Across the Boundaries: Extrapolation in Biology and Social Science*. Oxford: Oxford University Press.

Stigler, S. M. 1986. *The History of Statistics*. Cambridge, MA: Harvard University Press.

Sun, S.-C., P. A. Ganchi, D. W. Ballard, and W. C. Greene. 1993. "NF-κB Controls Expression of Inhibitor IκBα: Evidence for an Inducible Autoregulatory Pathway." *Science* 259:1912–15.

Tabery, J. 2004. "Synthesizing Activities and Interactions in the Concept of a Mechanism." *Philosophy of Science* 71:1–15.

2009. "Difference Mechanisms: Explaining Variation with Mechanisms." *Biology and Philosophy* (in press).

Taylor, J. R. 1997. *Introduction to Error Analysis: The Study of Uncertainties in Physical Measurements*. Sausalito, CA: University Science Books.

Trout, J. D. 1998. *Measuring the Intentional World: Realism, Naturalism, and Quantitative Methods in the Behavioral Sciences*. Oxford: Oxford University Press.

Waters, C. K. 2007. "Causes That Make a Difference." *The Journal of Philosophy* 104 (11):551–79.

Weber, M. 2005. *Philosophy of Experimental Biology*. Cambridge: Cambridge University Press.

Wimsatt, W. C. 2007. *Re-engineering Philosophy for Limited Beings: Piecewise Approximations to Reality*. Cambridge, MA: Harvard University Press.

Woodward, J. 2002. "What Is a Mechanism? A Counterfactual Account." *Philosophy of Science* 69:S366–S77.

    2003. *Making Things Happen: A Theory of Causal Explanation*. Oxford: Oxford University Press.

    2008. "Cause and Explanation in Psychiatry: An Interventionist Perspective." In *Philosophical Issues in Psychiatry: Explanation, Phenomenology and Nosology*, ed. K. Kendler and J. Parnas. Baltimore: Johns Hopkins University Press.

Ylikoski, P. 2013. "Causal and Constitutive Explanation Compared." *Erkenntnis* 78:277–97.

Cambridge Elements ☰

Philosophy of Biology

Grant Ramsey

KU Leuven

Grant Ramsey is a BOFZAP Research Professor at the Institute of Philosophy, KU Leuven, Belgium. His work centers on philosophical problems at the foundation of evolutionary biology. He has been awarded the Popper Prize twice for his work in this area. He also publishes in the philosophy of animal behavior, human nature, and the moral emotions. He runs the Ramsey Lab (theramseylab.org), a highly collaborative research group focused on issues in the philosophy of the life sciences.

Michael Ruse

Florida State University

Michael Ruse is the Lucyle T. Werkmeister Professor of Philosophy and the Director of the Program in the History and Philosophy of Science at Florida State University. He is Professor Emeritus at the University of Guelph, in Ontario, Canada. He is a former Guggenheim fellow and Gifford lecturer. He is the author or editor of over sixty books, most recently *Darwinism as Religion: What Literature Tells Us about Evolution; On Purpose; The Problem of War: Darwinism, Christianity, and their Battle to Understand Human Conflict;* and *A Meaning to Life.*

About the Series

This Cambridge Elements series provides concise and structured introductions to all of the central topics in the philosophy of biology. Contributors to the series are cutting-edge researchers who offer balanced, comprehensive coverage of multiple perspectives, while also developing new ideas and arguments from a unique viewpoint.

Cambridge Elements$^{\equiv}$

Philosophy of Biology

Elements in the Series

The Biology of Art
Richard A. Richards

The Darwinian Revolution
Michael Ruse

Mechanisms in Molecular Biology
Tudor M. Baetu

A full series listing is available at: www.cambridge.org/EPBY